COLOR, TEXTURE & CASTING FOR JEWELERS

Library of Congress Cataloging-in-Publication Data

Codina, Carles.
[Aula de joyería. English]
Carles Codina's color, texture & casting for jewelers : hands-on demonstrations & practical applications. -- 1st ed.
p. cm.
Includes index.
ISBN 978-1-4547-0017-3 (pbk. : alk. paper)
1. Jewelry making. 2. Metal castings. 3. Color in design. I. Title. II. Title: Color, texture & casting for jewlers.
TT212.C59813 2010
739.27--dc22
2010037977

10 9 8 7 6 5 4 3 2 1

First published in English in 2011 by
Lark Crafts
An Imprint of Sterling Publishing Co., Inc.
387 Park Avenue South, New York, NY 10016

If you have questions or comments about this book, please contact:
Lark Crafts
67 Broadway
Asheville, NC 28801
828-253-0467

Manufactured in China

ISBN 13: 978-1-4547-0017-3

For information about custom editions, special sales, premium and corporate purchases, please contact Sterling Special Sales Department at 800-805-5489 or specialsales@sterlingpub.com.

For information about desk and examination copies available to college and university professors, requests must be submitted to academic@larkbooks.com. Our complete policy can be found at www.larkcrafts.com.

Project and Production Parramon Ediciones, S.A.
Text Carles Codina i Armengol
Photography Nos & Soto
Illustrations Farrés il-lustració editorial
Translation from Spanish Eric A. Bye, M.A.

First edition: February 2009

Rosello i Porcel, 21 9th Floor
08016 Barcelona (Spain)

A Company of the Norma Group
Of Latin America
www.parramonforeignrights.com
ISBN: 978-84-342-3380-5

Carles Codina. Pendants made from white and yellow sheet gold.

COLOR, TEXTURE & CASTING FOR JEWELERS

Hands-On Demonstrations & Practical Applications

Carles Codina

An Imprint of Sterling Publishing, Co., Inc.
New York
www.larkcrafts.com

Part 1: Color, Textures, and Finishes

Part 2: Modeling and Casting

PART 1

COLOR, TEXTURES, AND FINISHES

Mokume Gane, Keum Boo, Granulation, and Patinas

Introduction

To understand jewelry making, it is not adequate to know it on a theoretical level. It is essential and necessary to practice it directly on the workbench. Such is the case with the techniques of granulation, keum boo, and mokume gane. Replicating these processes in the traditional manner allows you to come into direct contact with the material and fully appreciate its construction requirements. This firsthand learning helps us understand that there is nothing random in ancient processes. Classic decorations do not necessarily take their shape purely from aesthetics. Rather, their style is sometimes determined by the technique, the technology, and the material used.

There are an infinite number of skills you can learn through practice. These include the decoration used in keum boo, the distribution of colors in mokume gane, and the classic arrangement of the spheres in granulation. Part one of this book focuses directly on these three ancient, ornamental techniques. This is an invitation to work the way people did centuries ago in Asia, or millennia ago in the Mediterranean cultures. You will employ the same procedures, materials, and tools. You will also learn some secrets that have not been made public until a few years ago. We don't aim to imitate or copy the ancient aesthetic, but rather to use ancient processes to develop a current design and be inspired by contemporary artists who are doing the same.

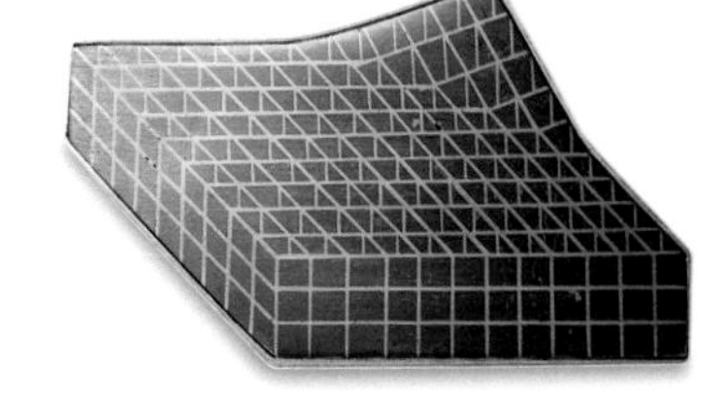

Hans Leicht. Brooch with fine gold wire inlay on a sheet of oxidized silver

It is difficult to appreciate contemporary European jewelry without the presence of the jeweler Hans Leicht. His contributions to alloys and metal combinations were numerous. As heirs and now orphans of his knowledge, who have enjoyed his company in learning jewelry making, we now wish to share it worldwide. Thank you, master.

Carles Codina i Armengol

Keum Boo

Keum boo is an ancient decorative technique of Korean origin that was used to embellish objects of daily and ceremonial use, generally tableware, earthenware bowls for food, and cutlery. These were decorated with various shapes as omens for their owner's good health. The gold decorations were applied inside items which would come into direct contact with the food, because they believed that the food took on the properties of the metal. Keum boo involves attaching fine sheets of 24-karat gold onto sterling silver or fine silver without using solder, thereby giving the objects a special beauty. It is necessary to prepare the silver in advance, along with the different gold motifs that will be used on the piece.

Clara Inés Arana. Brooches made using sterling silver bleached and oxidized with applications of the keum boo technique, pearl, onyx

Carles Codina. Pendant of oxidized silver and photogravure with inside application of keum boo

Clara Inés Arana. Sterling silver brooches and earrings with applications of keum boo, onyx, and 18- and 22-karat gold

Clara Inés Arana. Sterling silver pin with keum boo applications

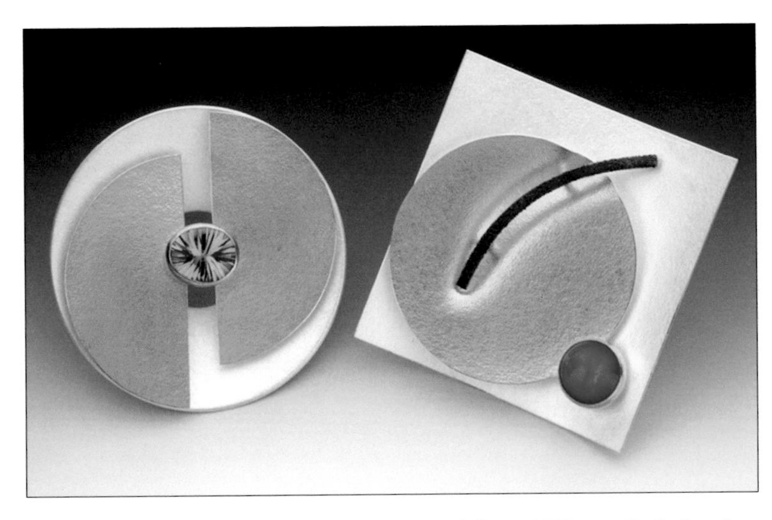

Clara Inés Arana. Sterling silver brooches with applications of silver and 24-karat gold using keum boo techniques, 22-karat gold, enamel, citrine, tourmaline

Clara Inés Arana. Oxidized silver brooch with keum boo applications, peridot

Clara Inés Arana. Silver and fine gold rings with pearls

Preparing the Gold Sheet

To begin, a sheet of 24-karat gold is made as thin as possible. The minimum thickness should be between 0.0004 and 0.0008 inch (0.01 and 0.02 mm), although it is possible to work with thicker metal. You should not use conventional gold leaf (that is used to decorate wood) because it will disappear completely when heat is applied.

The silver support and the thin gold sheet must be absolutely clean. Any particle of dust or impurity will keep the gold from adhering to the silver and will spoil the procedure. The metals must be prepared in an area free from dust and grease. The sheets have to be handled with fine, clean tweezers, keeping the fingers away from direct contact with the metal.

As shown in this demonstration, the fine gold is rolled in the rolling mill to produce the thinnest sheet possible. The resulting sheet is annealed and rolled further, but this time it is placed between two sheets of copper to reduce it even more.

1/ Take a sheet of fine gold and roll it in the rolling mill to produce the greatest reduction possible; then anneal it and degrease it in alcohol. The gold and copper sheets must be scrubbed with a brass brush and sodium bicarbonate, and then rinsed in distilled water.

2/ Once the sheet is annealed it is placed between two sheets of copper, and they are rolled together.

3/ The rolling process continues and produces a thickness of 0.0004 to 0.0008 inch (0.01 to 0.02 mm).

4/ After rolling, the gold sheet is annealed and cleaned. It is kept in alcohol while the piece to which it will be applied is prepared. It may be possible to make the sheet even thinner by rolling the gold between two sheets of parchment paper.

5/ To store the gold sheet, it's advisable to place it in a fold of parchment paper and put it between the pages of a book to keep it from getting dirty or wrinkled.

Cutting the Gold Sheet

The fine gold sheet is extremely fragile, and it must be treated with the utmost care. It is very soft, so it can be cut with the tip of a craft knife or with scissors. It is very important that it not be creased or wrinkled, and especially, that it not become dusty or greasy while being handled. Once the various shapes have been cut out they are kept in alcohol while the silver piece is being prepared.

1/ Any punch for school or hobby use can be used as a die for cutting out repetitive designs and shapes.

2/ Paper-cutting scissors can be used for cutting out various shapes repeatedly and for giving form to different designs.

3/ The purpose of the parchment paper is to avoid direct contact between the gold and the fingers, thus maintaining perfect cleanliness on the surface of the metal.

4/ If you wish to cut out a certain design precisely without touching the metal, place the gold between two sheets of parchment paper, and draw the design on the top layer of paper. Then, cut out the drawing with the tip of a hobby knife so that the cut carries through to the gold sheet.

5/ Once the shape is cut out it's a good idea to isolate the gold sheet by protecting it between two sheets of parchment paper, and keeping inside a thick book to keep it from wrinkling.

Surface Preparation

The thin gold can be applied to a silver sheet before it is soldered or it can be applied to a fully finished piece. The second option is preferable, because soldering after applying keum boo may produce air bubbles between the gold and the silver. You can solve this problem by pressing the gold with a burnisher while it's hot. The gold is applied directly onto the silver as long as it has a smooth, uniform surface and is prepared as shown in this demonstration.

It is necessary to produce a fine, spongy layer of fine silver on the base surface. In this process, called depletion gilding, the sterling silver item is repeatedly subjected to oxidation and cleaning, via pickle or acid. Oxidation draws the copper to the silver surface, and the acid removes it. The new fine-silver layer permits the adhesion of the fine gold sheet through heat and pressure. The result is a perfect adhesion that is impossible to remove by mechanical means and is able to withstand soldering work.

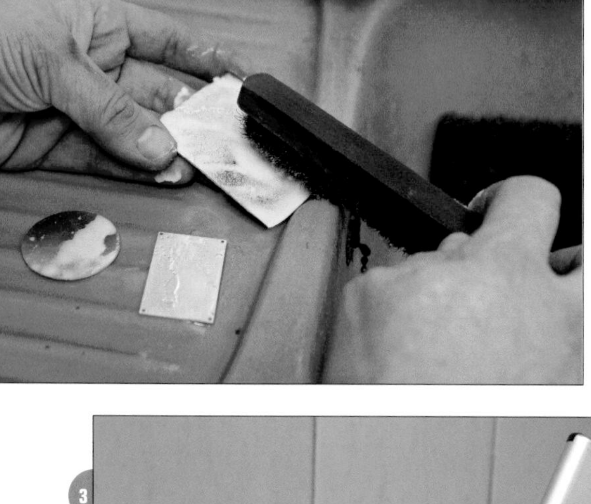

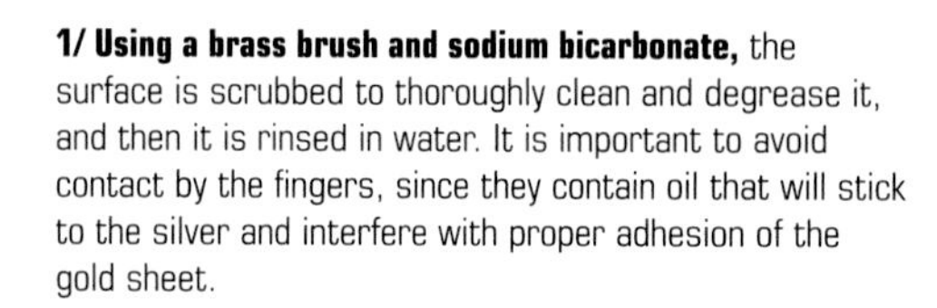

1/ Using a brass brush and sodium bicarbonate, the surface is scrubbed to thoroughly clean and degrease it, and then it is rinsed in water. It is important to avoid contact by the fingers, since they contain oil that will stick to the silver and interfere with proper adhesion of the gold sheet.

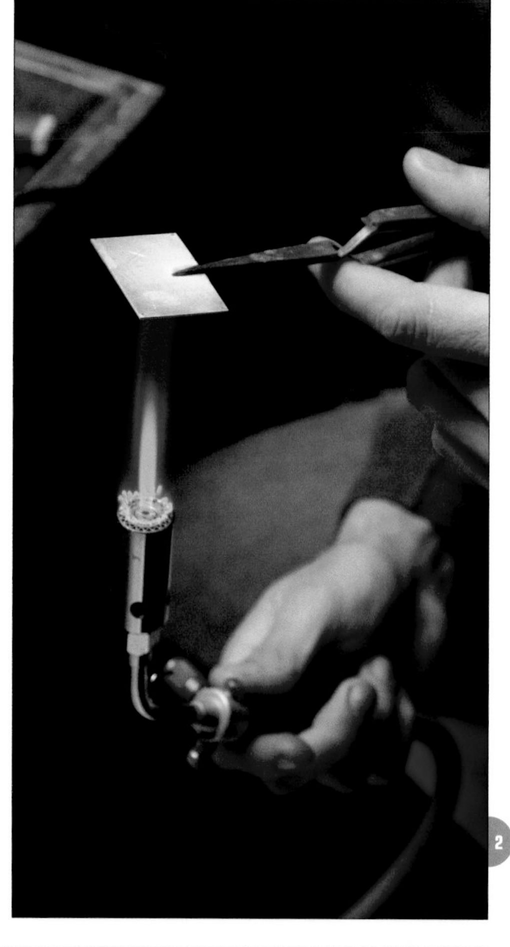

2/ The piece or the sheet is heated from the back, but without reaching annealing temperature; the oxidation merely needs to appear on the surface. Then it is put into pickle or acid, and this process is repeated eight to ten times.

3/ The pickle or acid must be new and have a strong concentration. It is essential to dry the piece after rinsing it in water; avoid touching the surface and disturbing the layer of fine silver produced on the surface.

4/ If the procedure is done correctly, when the succession of heating and pickling is finished, the piece will take on a white color because of the appearance of fine silver on the surface.

Application

The gold sheet can be applied directly onto the silver. The silver sheet is heated with the torch, and the fine layer of gold is placed on the surface using fine tweezers.

For more precise work, it is better to use a 2000-watt electric hotplate as shown in this demonstration. This common kitchen burner makes it possible to work comfortably since it warms the silver sheet to the right temperature, and leaves both hands free to apply the gold. Also, because this is an electrical device, it lessens any oxidation of the silver that later could cause an improper bond with the gold.

Finish

If the gold has been attached properly, the piece will not have oxidized, and it will not be necessary to pickle it. It can be left as is, or it can be textured by putting the keum boo sheet through the rolling mill in combination with a piece of textured paper or emery paper. The goal is to create fine textures. A pronounced texture could break the gold sheet. The gold can also be polished if you work gently.

1/ The sheet of silver is placed on the burner so it reaches the proper temperature, about 750° F (400° C). Then, using fine tweezers, the gold sheet is placed in position on the warm silver sheet.

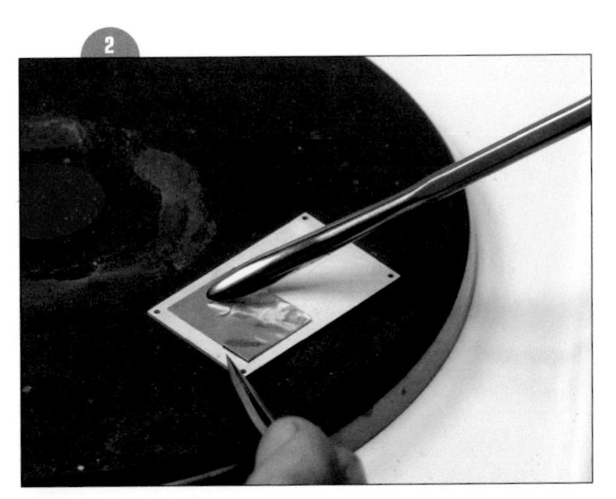

2/ A burnishing tool moistened with water is used to press the gold sheet onto the silver, spreading it out evenly toward the edges, without creating wrinkles and bubbles. At this point there should be a feeling of roughness as you spread out the gold sheet; it indicates that the two metals are adhering properly.

3/ Once the gold sheet is spread out, the entire surface is burnished to ensure a proper bond. It is essential to moisten the steel burnisher in water repeatedly to keep the tool from becoming overheated.

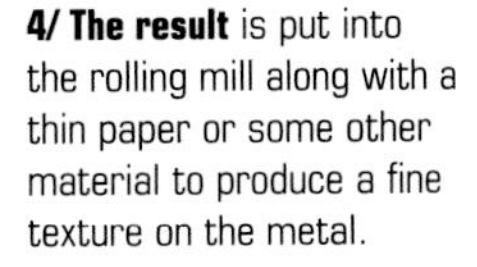

4/ The result is put into the rolling mill along with a thin paper or some other material to produce a fine texture on the metal.

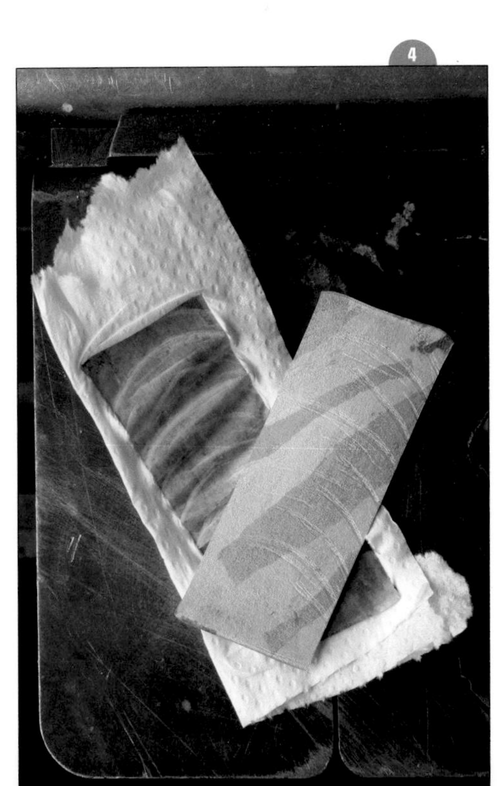

Plating

This is another jewelry technique that makes it possible to combine different colors of silver and gold. Plating goes by different names based on the country or area where it is done. The following project is a quick variant of traditional plating, and it can be done with few tools. The technique allows the use of any precious metal, such as fine gold or pure gold, palladium, white gold, or silver. The metal bond is produced at a higher temperature than with keum boo, so the bond is stronger. Since there is a true fusion of the surfaces, subsequent work such as soldering can be done. However, this plating process requires working with slightly thicker gold sheets, which are more expensive.

The most practical method involves preparing two steel sheets about 5 mm thick and previously scaled. To scale a sheet, it is heated and quickly cooled in water, and then the surfaces are coated with graphite or red ochre to keep the metal from sticking to them when it melts.

Miquel Gasso. Ensemble consisting of a bracelet, ring, and pendant in yellow and white gold made with the plating technique

Miquel Gasso. Silver and pure gold pendant

Marta Sánchez. Oxidized silver ring with pure gold applications

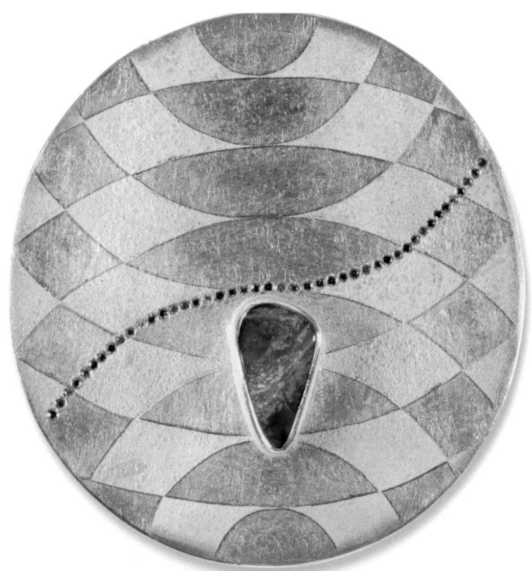

Michael Zobel. Papageno pendant brooch made from 18- and 24-karat gold with a 14.27-carat opal doublet and various colored diamonds. Design: Atelier Zobel. Photography: Fred Thomas

1/ In this demonstration, a 1-mm-thick sterling silver sheet and a very thin sheet of pure gold were used, but this project can be done with any other precious metal. The two metals to be joined must be moistened with flux and the box must be held together firmly with steel binding wire.

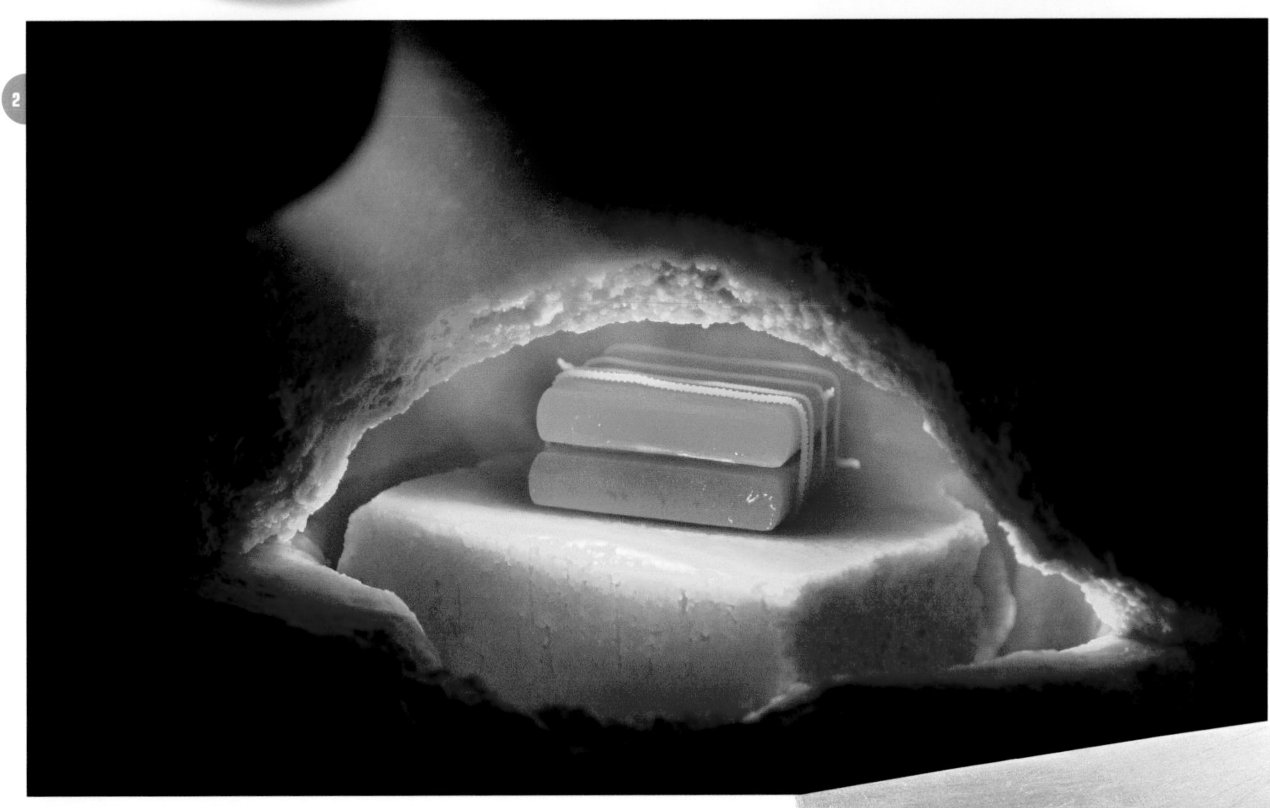

2/ The furnace is preheated. The sheets are placed inside, and the temperature is increased to produce the fusion of the gold and silver. When the fusion is done, everything is quickly taken out and allowed to sit before removing the oxidation in pickle solution.

3/ The process produces a fusion between the two metals. Once cleaned, any kind of work can be done with the sheet since there is no solder between the two metals.

4/ For this project, the fused metal is layered with a piece of coarse sandpaper, and then placed between two copper sheets to protect the cylinders of the rolling mill.

5/ For the following brooch, a setting was made for the tourmaline, using a beveled 0.8-mm sheet joined with hard solder.

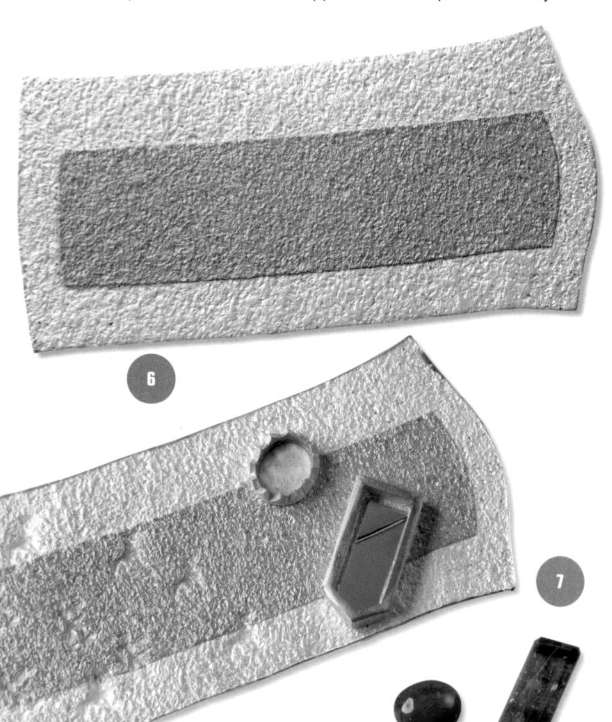

6/ The metal set was put through the rolling mill, and the texture was imprinted. Now it is necessary to anneal the metal and pickle it again.

7/ A setting was made for the ruby using 0.6-mm gold wire. The settings were soldered to the sheet, and the clasp was made.

8/ The finished brooch with the gemstones properly set.

Melted Filings

This is a simple procedure that involves melting precious metal filings on a sheet of another precious metal. Precious metals of contrasting colors must always be selected. For this demonstration, a silver base and 18-karat gold filings were used.

The filings must be perfectly clean. A piece of pure gold is filed with a coarse file. A magnet is used to remove any steel remnants. The gold is burned in an old frying pan, and then it is put into sulfuric acid to remove the oxide. The gold is rinsed in water and dried. Depending on the coarseness of the file used, different sizes of filings are produced. It is also possible to create variations and different effects by melting the filings onto the base sheet to a greater or lesser degree.

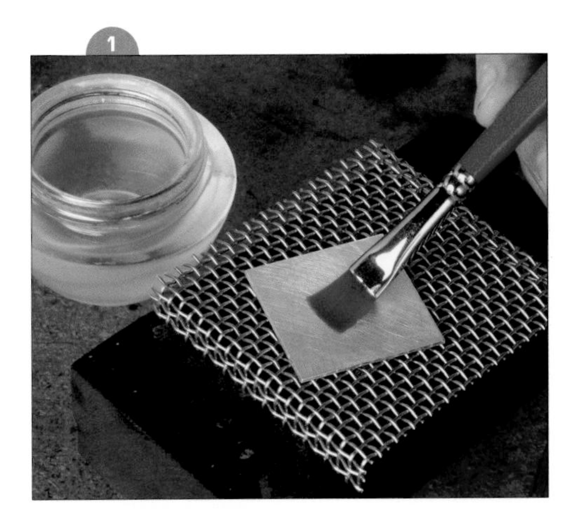

1/ A sterling silver sheet about 1-mm thick is placed onto a steel grate and coated with liquid flux.

2/ The liquid flux is heated gently until dry.

3/ The yellow-gold filings are sprinkled evenly onto the silver sheet.

4/ Heat is applied to the sheet with the torch until the filings melt slightly on the surface.

5/ To produce an interesting texture, the sheet was put through the rolling mill along with a piece of paper towel. Then it was annealed again and pickled.

6/ Two disks were cut with a disk cutter. On one disk, the silver was blackened to emphasize the gold filings.

Mokume Gane

Mokume gane is a technique that was used in Japan to decorate the hilts of Samurai swords in the sixteenth and seventeenth centuries. These decorations generally were an indication of rank and social status. Translated, *mokume* means vein, and *gane*, metal. This is precisely the effect sought by this ancient technique.

The process involves creating a block made up of different sheets of metal without using solder. It makes use of the ability of metals to bond with one another through diffusion. This occurs when the metals are subjected to a certain pressure and temperature in a reducing atmosphere. Under these special conditions, the result is a slight surface fusion that allows the various alloys that make up the metals up to bond with one another. Once that bond is formed, the result is a compact block composed of different metal alloys that can be worked much like an ingot.

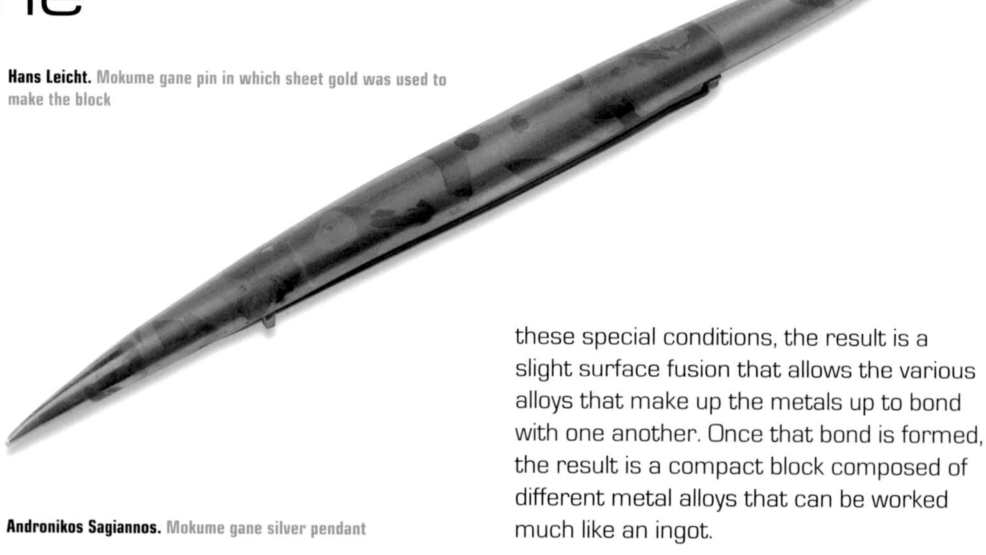

Hans Leicht. Mokume gane pin in which sheet gold was used to make the block

Andronikos Sagiannos. Mokume gane silver pendant

Alloys

The standard alloys for mokume gane are Shakudo, Shibuichi, and Kuro Shibuichi, but interesting results can also be achieved with gold, fine or sterling silver, and even bronze. It is important to use alloys that have malleability and melting points that are close to one another. This is why copper and its alloys are used as a base metal. The copper content in the alloy improves its malleability. It is preferable and advisable to bond copper alloys to silver alloys, and not with bronze alloys.

When you want to bond alloys with widely different melting points, the procedure is done separately. Gold, silver, and copper start to bond around 1355°F (735°C), and on average they melt at around 1454°F (790°C). Brass or bronze alloys bond with copper between 1490°F (810°C) and 1796°F (980°C). It is not advisable to use brass with silver, since it produces an alloy with a low melting point that can spoil the block. Specialists in the technique can even join sheets of stainless steel, titanium, and aluminum, and achieve astonishing colorations.

Three or more alloys can be used to make the block, but experience indicates that the best results are obtained by mixing successive layers of two alloys that contrast well, for example Shakudo with fine silver or gold, or Shibuichi with copper or Shakudo.

Kuro Shibuichi is made by fusing different proportions of Shakudo and Shibuichi. Because of its arsenic content, Kuromi Do is dangerous to make. In this chapter, pure copper has been substituted.

Hans Leicht. Mokume gane bracelet made from a block from which various projects were made

Hans Leicht. Mokume gane earrings

Hans Leicht. Pendant made with successive perforations on a block of mokume gane

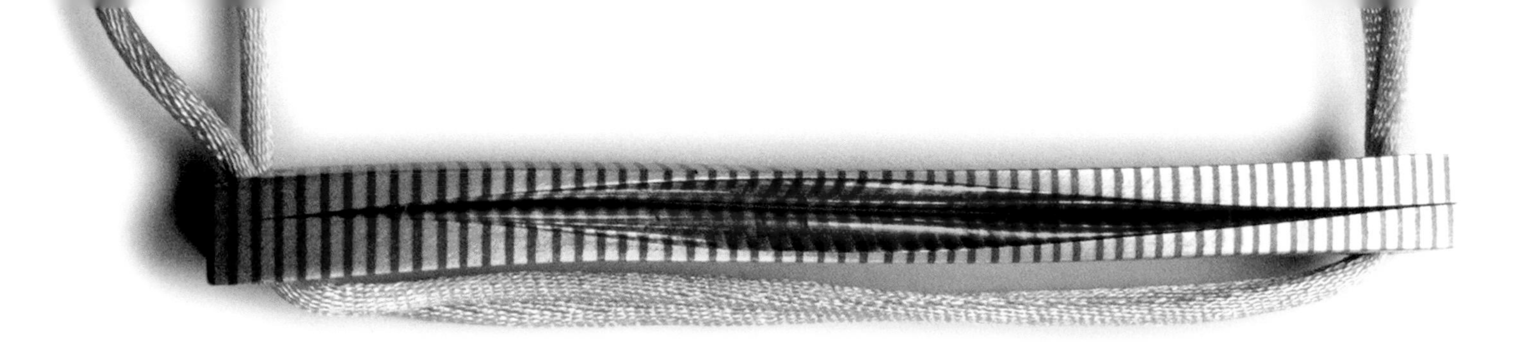

Simon Viktória. Mokume gane brooch made from two pieces of block screwed together

The Simplest Mokume

As an introduction to this technique, the simplest mokume gane is made with sheets of fine silver and copper. This combination offers good malleability and a good visual contrast between sheets with a low cost for materials. Another valid option involves preparing a block from sheets of Shakudo and fine silver. Once your technique is perfected, you can introduce a new alloy or a sheet of fine gold in the upper layers of the block.

Metal Melting Temperatures

Metal	Temperature
Fine Silver	1760° F (960° C)
Sterling Silver	1639° F (893° C)
Fine Gold	1944° F (1062° C)
Pure Gold	1661° F (905° C)
Shakudo	1922° F (1050° C)
Shibuichi	1742° F (950° C)
Shiro Shibuichi	1562° F (850° C)
Kuro Shibuichi	1814° F (990° C)
Kuromi Do	1958° F (1070° C)
Copper	1980° F (1082° C)
German Silver	1960° F (1071° C)
Aluminum Brass	1749° F (953.8° C)
Nickel Silver	2030° F (1110° C)

Two basic alloys are Shakudo and Shibuichi. With them, it is possible to make Kuro Shibuichi. The alloys must be laminated, and the metal must be kept from fracturing or developing small pores or surface imperfections.

Shakudo
4.8% fine gold and 95.2% copper
2.5% fine gold and 97.5% copper

Shibuichi
40% fine silver and 60% copper
30% fine silver and 70% copper
5% fine silver and 95% copper
3% fine silver and 97% copper

Shiro Shibuichi
60% fine silver and 40% copper

Kuro Shibuichi
83.3% Shakudo and 16.7% Shibuichi
71.4% Shakudo and 28.6% Shibuichi
58.8% Shakudo and 41.2% Shibuichi

Preparing the Sheets

Mokume gane requires the utmost care in all its phases, but it is during the preparation of the metal sheets that special attention must be paid to polishing and cleanliness. The sheets must end up 1 mm thick, the rough edges must be removed, and the metal must be subjected to a succession of emery papers to produce a very fine grain. Then, the metal is rubbed with pumice powder or a pumice stone and gone over with steel wool. This way, the surfaces are perfectly flat and free from dust and grease. The sheets are then put into equal parts of nitric acid and water, they are rinsed in water, and then they are kept in vinegar or alcohol to avoid oxidation and contamination. It is very important to avoid any contact with dust and oils from the fingers, since this would produce a faulty joint and completely spoil the bond in the block.

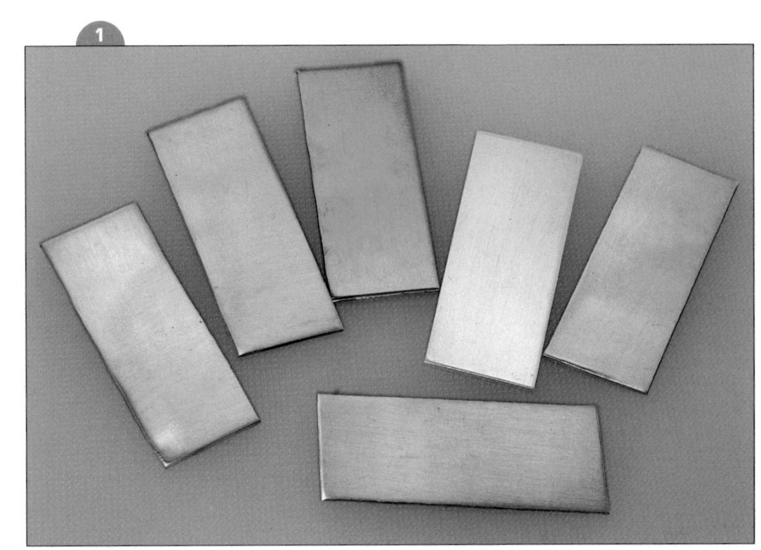

1/ The sheets are cut to the size of the box. After annealing them, the sheets are flattened properly and a progressive polishing is begun to produce a surface free of impurities.

2/ Once the sheets are flattened, they are rubbed in water using a cut pumice stone and emery powder to produce a uniform surface.

3/ Next the sheets are rubbed with a silicone-and-emery chamois to improve the surface and leave it finely polished and regular.

4/ The sheets are rinsed in water and quickly put into denatured alcohol where they will remain until it's time to put them into the steel box.

5/ Once clean, the sheets are removed from the alcohol and put into a refined borax solution, avoiding direct contact between the metal surface and the fingers.

Making the Box

A box is made from iron or refractory steel, and several designs are possible. The box must be capable of distributing heat uniformly, and the covers must be thick enough to allow the compression of the sheets that make up the block when the melting point is reached.

1/ To construct the box, a piece of pipe and a couple of steel shapes at least 5 mm thick are cut out. The top part of the pipe is cut off, and two pieces of rail are fitted in the center of the box.

2/ The box is first scaled (black oxide scales are made to appear on the surface), and the tensions inside the metal are eliminated. To do this, the box is heated to red and quickly quenched in water. This is done two or three times.

3/ The purpose of scaling is to stabilize the steel and produce a layer of scale on it to keep the block from sticking to the box during the process of bonding the sheets.

4/ Before putting the sheets in the box, the interior surfaces are painted with protective graphite or with diluted ochre earth.

5/ The sheets are removed from the solution of soldering fluid and distilled water and arranged by holding them by the ends. Avoid touching the surface with the fingers.

6/ The box is bound with thick welding wire to keep the sheets from moving during the melting process.

Another Type of Box

A different box design is made for producing larger blocks. It requires two steel plates, each 8 mm thick. Holes are drilled in each corner so the bolts fit in them loosely. This extra space is necessary so the weight of the top plate will press down on the block when the box is inside the furnace.

1/ To scale the box, it is heated bright red and then plunged into a bucket of cold water.

2/ The box is scaled several times to stabilize it and bring all the steel scale to the surface. The inside surface is protected with graphite or ochre.

3/ The sheets are put inside the box in the proper order and compressed using the appropriate nuts.

4/ This type of box accommodates more than ten sheets of larger dimensions than the previous box design. They are placed inside and compressed forcefully. Then, the box is put into the preheated furnace.

Preparing the Furnace

A furnace is made using refractory bricks. Charcoal and coal are put into it, the latter in smaller amounts, to create a reducing atmosphere. An environment with a high carbon content must be produced. This is necessary for success with mokume gane. It is possible to make the furnace using normal refractory bricks covered with a thermal blanket, or you can use a small refractory furnace of the type commonly used in many workshops. It is also convenient to do this procedure in a blacksmith's forge, using a mixture of charcoal and coal.

2/ A reducing atmosphere is produced when the coal has reduced its volume by around 50 percent. At this time, the box is put inside the furnace, and the door is shut.

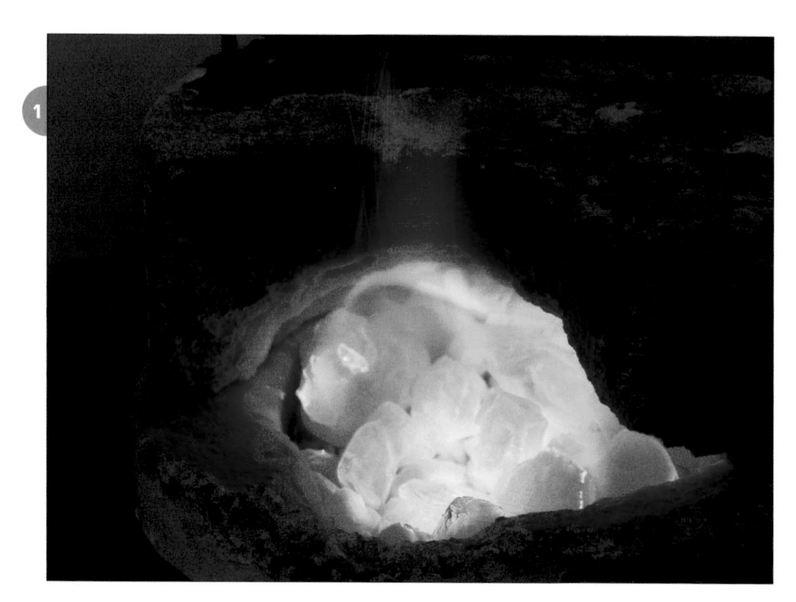

1/ The furnace must be preheated with the coal inside until it is glowing. It is important to continue loading the furnace with charcoal to produce the reducing atmosphere necessary for the proper bonding of the metal sheets.

3/ The temperature is increased with a propane torch. To reduce the risk of oxidation as much as possible, it is important that the flame be broad and reducing and to avoid excess pressure and oxygen.

4/ The crucial moment in every mokume gane process involves detecting a subtle glow among the various metal layers, indicating that they have bonded. At this precise instant, the block must be removed from the furnace immediately.

An Electric Furnace

Mokume gane can also be made in a closed electric furnace with a good heat control. The box and the sheets must be covered with charcoal to avoid oxidation. If an electric furnace is used, it is necessary to keep the box about 575°F (300°C) below the lowest alloy melting temperature, and raise it gradually until it matches the metal's melting temperature. A uniform temperature must be reached that completely penetrates the block and the box. This process can take between five and ten hours. When working in a closed furnace, it is advisable to do a few prior temperature tests.

Forging

Once the block it produced, it is etched in sulfuric acid and rinsed properly. Then, the forging begins to compact the block and reduce its size. Proper forging will keep the block from breaking excessively at the ends. It is preferable to forge hot, at a black heat. This involves first heating the block in the same furnace, and then starting to forge when the block loses its red color. Hold the block on the anvil with pliers. Forge it by striking repeatedly with a forging hammer on both faces of the block. This process is repeated several times to produce a 50-percent reduction in size. From this point on, the metal can be put through the rolling mill. Once the block has been reduced by more than half, it can be cut through the middle and joined once again inside the steel box. This doubles the number of layers in the block.

Ways to Work the Block

There are many creative possibilities, especially if a block is made with contrasting colors. Strips can be cut out and soldered together, a strip can be cut out and stretched to make a ring, or the strip can be forged to make a pendant or any other desired shape. In the following pages, we will see only the basic ways to work, but starting with these examples an infinite number of projects and items can be made.

1/ Alloys of silver and gold in particular must be forged at black heat, around 850°F (400°C). If they are forged when the metal is very hot, the block may break.

2/ The entire surface must be forged uniformly, including the sides. This way, the block will be compacted without areas of collapse or fractures between the sheets.

3/ A block made using various alloys of Shakudo, Shibuichi, and Kuro Shibuichi. To keep the fractures on the edges from spreading, it's a good idea to file the ends as the block is reduced through forging.

Cutting the Block

You can maintain the original size of the block or slightly reduce it with a jeweler's saw. The various laminates can be used to produce new ingots for making rings. They can also be soldered together to make new laminates.

1/ Once the block is produced it is annealed and forged to produce a 50-percent reduction. At this time, you can roll it with the rolling mill or begin any other work process. Depending on what your forged design requires.

2/ Some longitudinal pieces are marked and cut out of the block. They are filed and polished precisely. It is possible to do countless projects with these strips, depending on the colors and the work process selected.

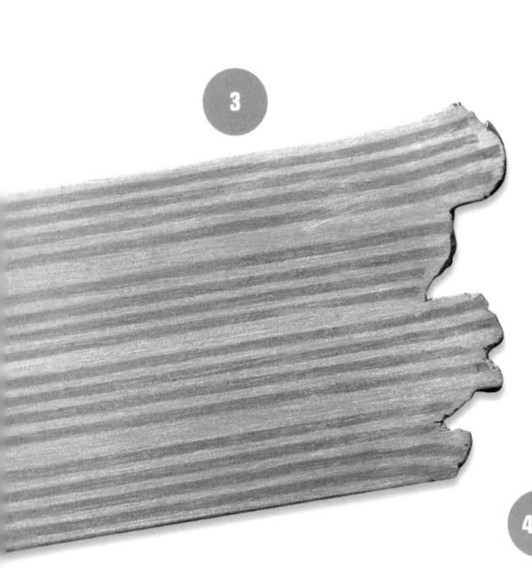

3/ As the photo shows, the pieces of metal can be joined on the sides and run through the rolling mill to create a new laminate for making new pieces.

4/ For this sample, it was decided to strike the side forcefully with a 5-mm steel rod, and drive it into the block to produce an interesting shape.

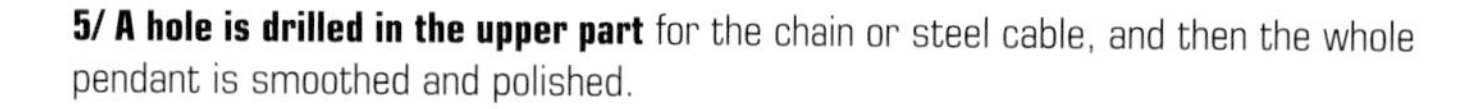

5/ A hole is drilled in the upper part for the chain or steel cable, and then the whole pendant is smoothed and polished.

A Laminated Plate Made from the Block

Hans Leicht produces an interesting metal plate based on a block composed of 13 different sheets. Once the block is produced, it is cut into squares that are again bonded together into a new block. This allows different construction possibilities. In this sample, we have decided to reduce the block by forging it to produce a sheet about 1 mm thick that will be used to make various pieces of jewelry.

1/ Thirteen sheets are prepared, seven of silver and six of copper or Skakudo. They are put into the box and compressed with the bolts that hold the box together.

2/ Once the sheets are bonded together through fusion in the furnace, the entire outer perimeter of the block is filed before starting the forging process.

3/ For this project, Hans Leicht marks out different squares on the surface of the block. He cuts them out with a saw to produce perfect cubes.

4/ Once the cubes are cut out, they are switched around to produce contrasting directions in the laminations. Once fitted together in a new order, they are soldered with hard solder to produce a new block with different effects.

5/ The block is forged on the anvil to produce a 50-percent reduction. Then, it is passed through the rolling mill to produce the desired thickness. The laminate is now ready for any kind of work.

Drilling

This process involves drilling a number of partial holes in a laminated block that is about 50 percent of its original thickness. Subsequently, additional holes are drilled using different-sized drill bits. After forging and a subsequent reduction in the rolling mill, the result is a sheet filled with interesting effects.

1/ A drill press is used to drill partial holes, just beyond the middle of the layers that make up the block.

2/ You can make holes of different sizes, or as in this example, make regular holes all the same size.

3/ You can also reduce the block in the rolling mill to half its thickness, and drill a number of holes with different-sized bits.

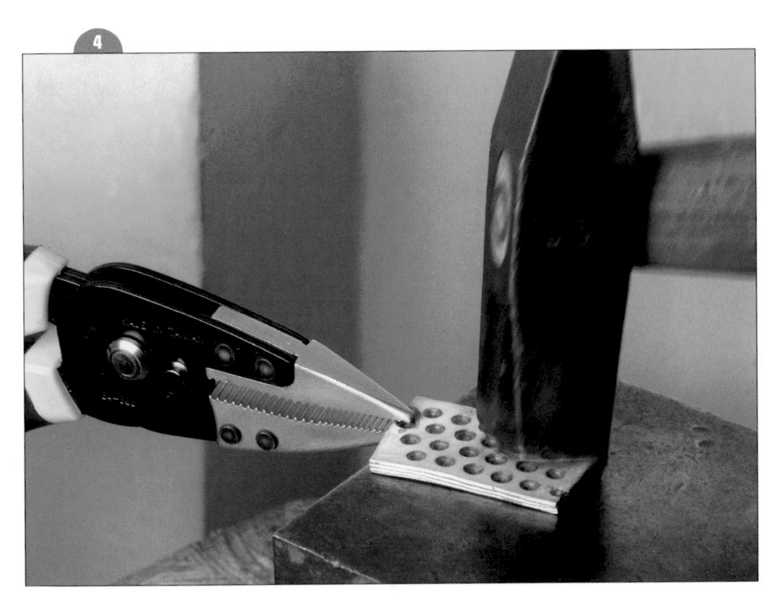

4/ The block is forged to produce a gradual and uniform reduction in the design. Once it is reduced 50 percent, the block is rolled through the rolling mill.

5/ Continue forging and making new holes until an interesting result is produced.

6/ Drilled mokume gane laminate is exposed to nitric acid to give the metal surface a little dimension.

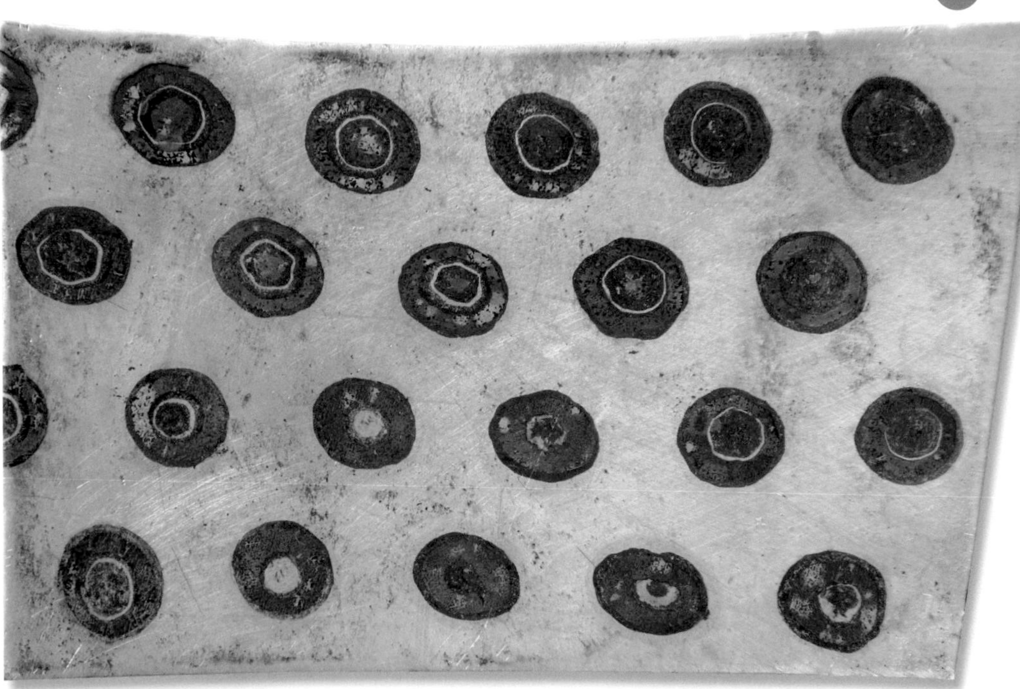

7/ Drilled mokume laminate given a heat patina.

Pressing

Hans Leicht has come up with a new method for working on a block of mokume gane. It involves a piece of steel that makes square holes. The result is a sheet of mokume gane with square designs that can be used for making jewelry.

1/ To drive the steel points inside the block, it is necessary to strike the surface with a heavy hammer and to make the points penetrate at least six or seven layers.

2/ The process requires several annealing procedures to drive the steel far enough into the block.

3/ This is the result of driving the piece of steel into the block.

4/ Once you reach the desired depth, you forge the block to reduce it by 50 percent.

5/ The block is annealed and run through the rolling mill to produce a sheet.

6/ The result of rolling the plate to a thickness of about 0.09 mm. Now it can be used to make a broad variety of jewelry items or used for various silversmithing projects. Since it contains no solder, it can be mounted in a turning lathe, it can be engraved, or it can be forged in the same way as a laminated silver sheet.

Filing

Another very common way to work with a mokume gane block involves cutting through various layers with different shapes of files. The purpose is to remove metal from the deepest layers. In subsequent forging and laminating, these layers will show on the outside as strips of color.

1/ You can file down to the deepest layers of the block, but it's a good idea to stop short of the last two or three layers.

2/ The piece is annealed and forged to reduce its volume by half. File any edges that may stick out from the metal. It helps to use a round file, especially for smoothing the top edges.

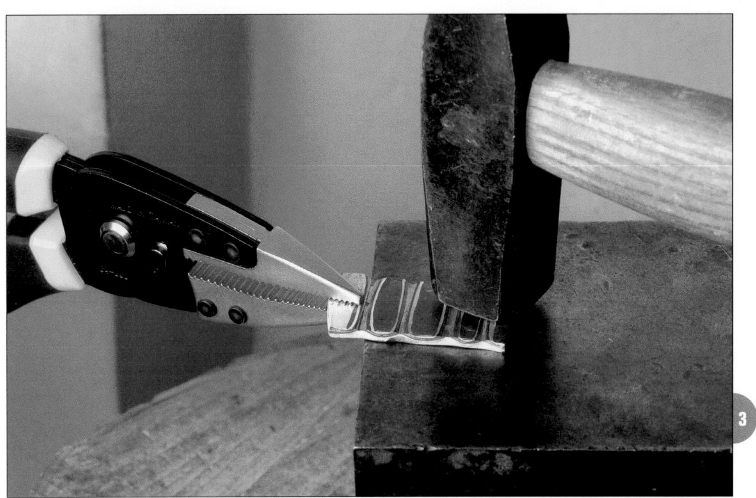

3/ Continue forging and annealing the block, always working hot at what is known as a black heat.

4/ Once you reach a reduction of 50 percent, laminate the block to produce a sheet that is the desired thickness for making a ring.

5/ Cut out a strip and bend it in the shape of a ring. Solder it and hammer it lightly. Polish the surface with a very fine emery paper, and the ring is ready for its final finish.

Punching

Some interesting effects can be created by forging or rolling a block to a thickness of about 0.6 mm, and then chiseling or striking the back side with cold chisels, punches, or hammers. Create lines, designs, or simply strike the metal with a spherical punch, and then file the distorted metal to produce the pattern.

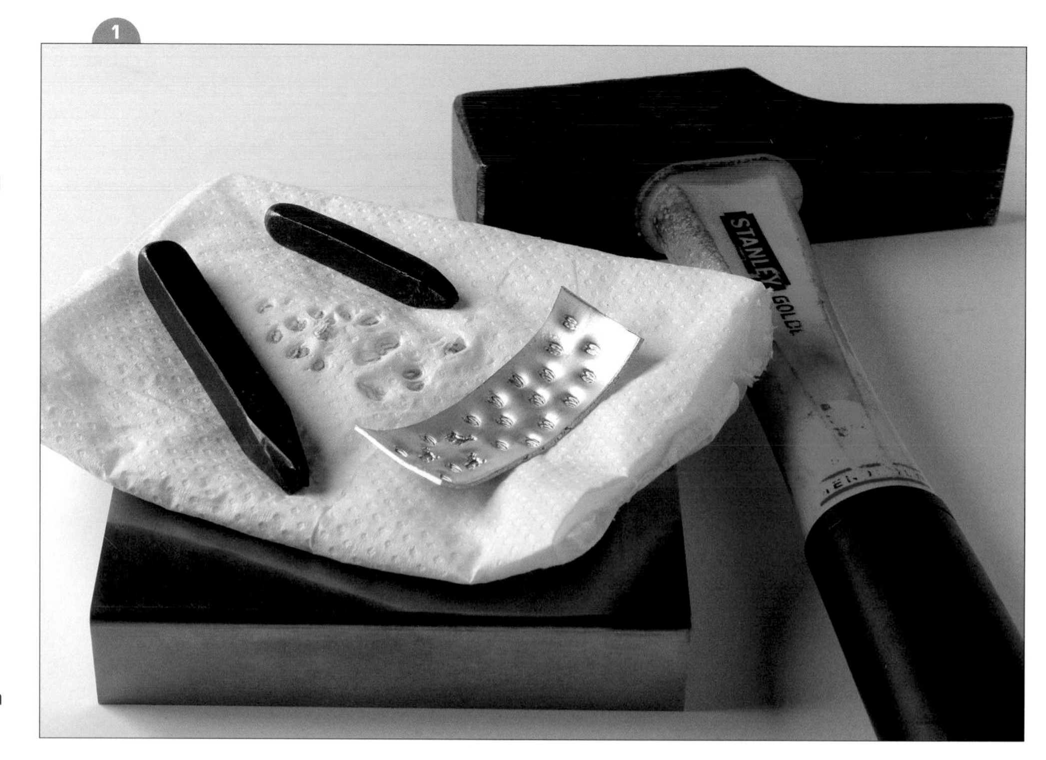

1/ Place a piece of soft paper between the mokume gane sheet and the steel anvil. Strike the back of the sheet with various steel punches.

2/ Before laminating, solder must be placed on the back of the sheet to fill the hollows created with the punch. This keeps the design from breaking during the rolling process.

3/ The distortion created with the punch is filed on the front side of the sheet, causing different layers to appear on the surface.

4/ Once the plate is polished, it is ready for making any kind of jewelry.

Final Coloring

This step is important because it makes the various metal layers stand out visually. Once the item is finished, the surface must be completely cleaned and degreased. It is filed and degreased with an agent such as trichloroethylene, or else it is boiled in water. Then, it is submerged in the boiling colorant and stirred for at least 10 minutes, at which time the color is checked. The piece is removed from the solution and rinsed in water. This process must be repeated until the desired effect is achieved. Once the piece is colored, it's a good idea to wax the surface with floor wax or car wax, and buff it with a soft cloth.

Rokusho

Rokusho is traditionally used to create different colorations. This compound colors Shakudo purplish-black. The greater the gold content in the alloy, the brighter the purple color produced. Copper will take on a reddish-brown color. Shibuichi, Shiro Shibuichi, and nickel silver take on colors of light gray. Gold and silver are not affected. They retain their original color.

Rokusho can be prepared with the following formula:

about 92 grains weight (6 g) copper
23 grains weight (1.5 g) copper sulfate
23 grains weight (1.5 g) table salt
1 liter distilled water

Another colorant similar to Rokusho can be prepared by mixing 92 grains weight (6 g) of copper sulfate with 15 grains weight (1 g) of salt and 30 fluid ounces (900 ml) of distilled water.

A mixture of 90 grains weight (6 g) of copper sulfate and 15 grains weight (1 g) of common salt is prepared by dissolving them in 30 fluid ounces (900 ml) of distilled water. Return to a boil and put the rings in the solution for about 20 minutes.

Result after 10 minutes of boiling

Result after 30 minutes of boiling. When the patina is done, the surface is waxed with a little pure wax.

Another ring made with mokume gane and patinated with the same formula, but gold was used on the outer surface.

Granulation

Granulation is an ancient decorative technique that involves joining small granules or balls of gold or silver together or on a precious metal surface. Before soldering was used as a means of joining metals, fusion through granulation was used. With this technique, it was possible to join minute pieces of fine or high-grade gold together and make a great variety of decorations. Used along with engraving, stamping, and filigree, granulation ended up defining a whole aesthetic style.

History

The first granulation works found were in the royal tombs of Ur, dating from the year 2500 BC. These were granules of great quality and finish. that measured about 2 mm and were perfectly attached. It is also known that granules measuring 0.4 to 1.1 mm were made in Troy. Delicate works have also been found in Crete and ancient Egypt as well as in the whole Phoenician culture, which spread this type of decoration throughout the Mediterranean region.

The highpoint of the granulation technique was reached during the eighth and seventh centuries BC, in what we might term the goldsmithing of pre-Hellenic Greece. Significant technical progress was made that affected not only granulation but also filigree, repoussé, and embossing, and which generated a true decorative style. Granulation appeared in Etruria in the eighth century, imported by Phoenician

Granulated gold rings with niello applications.

Rings made from old, commercially manufactured chains give discarded materials a new use. To accomplish this, Tensi Solsona used a new formulation of granulation paste.

merchants. It reached its high point in the sixth century, in which granulation work never before seen was produced, with granules of 0.1 to 0.2 mm. The technique penetrated into southern Spain as a result of importation as did the complementary application of filigree. It is known that granulation also reached China in the third century AD, and precious works made with this technique have been found there.

Similar procedures for bonding metals without using solder have also been found in pre-Columbian jewelry, in the Calima and Taironas cultures, in Colombia, and in the Mochica and Chimú cultures in present-day Peru.

In a primitive manner, copper was diluted in vinegar to produce an acetate solution, and organic natural adhesives were added. After firing in an atmosphere free of oxygen, copper was produced. By increasing the temperature in a furnace or with a torch, the copper melted a little sooner than the fine gold and this produced a micro-solder. This was a totally intuitive process that varied from one culture to another, but surface decorations of great beauty were created.

Pendant with granulated spheres in silver with Mares stone by Dani Fábregues

Metal Granules

Joining together different elements of pure metal without using a metal with a lower melting point or solder requires the addition of copper. This is applied by means of a paste compound. In ancient granulation, this granular paste was composed of ground malachite and organic glue.

Although it is possible to granulate 18-karat gold or sterling silver, better results are achieved with fine silver or 20- to 22-karat gold. In the following pages, various types of bonds with granular paste will be made. We will use fine gold alloyed with copper, as well as small granules of fine silver. It is preferable for the metal alloy to have a high melting point, because it is easier to control the temperature, and the resulting granules are more clearly defined. This also gives the particular luminousity and smooth finish of ancient granulation.

In the Furnace

When it changes to a liquid state, metal tends to occupy the smallest possible space. Because of this, when a small sheet or ring is melted, it changes into a small sphere or granule.

A furnace is used when it is necessary to make a large number of very small granules. In reality, this is a very ancient method for making granules. A steel refractory cylinder or a graphite crucible is used. A first, medium-thick layer of charcoal is placed in the crucible, followed by a layer of very fine graphite. Next, various small pieces of metal are cut out or small rings are prepared. They are spread out evenly on the layer of graphite. The metal is covered with new layers of graphite. This process is repeated, adding three or four layers until the cylinder or crucible is completely filled. It is put very carefully into a furnace preheated to 2200°F (1200°C) for gold or 1830°F (1000°C) for silver for a minimum of two hours. After that time, it is allowed to cool, and the whole assembly is put into water. The charcoal and the graphite are poured off, and the small spherical granules of gold or silver remain on the bottom.

1/ To make a large number of small granules, a very thin silver wire is drawn and folded onto itself several times. It is then cut into small pieces using snips.

2/ The charcoal is ground in a mortar and the graphite is prepared. It can be bought in the form of a finely ground powder.

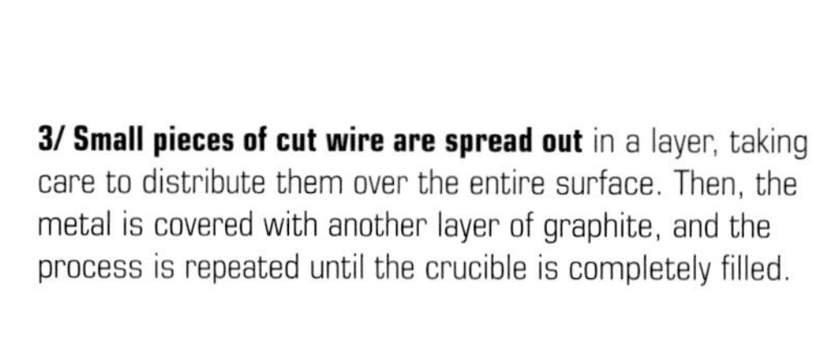

3/ Small pieces of cut wire are spread out in a layer, taking care to distribute them over the entire surface. Then, the metal is covered with another layer of graphite, and the process is repeated until the crucible is completely filled.

4/ Once they are pickled and dried, the granules that are not sufficiently round are removed.

5/ A sieve for stones is used to separate the spheres by size.

6/ The spheres are classified separately by size, and the defective granules are removed.

One Way to Load Copper

This is a quick way to load copper onto the granules. It is not as effective as granulation paste, but it is useful for small quantities.

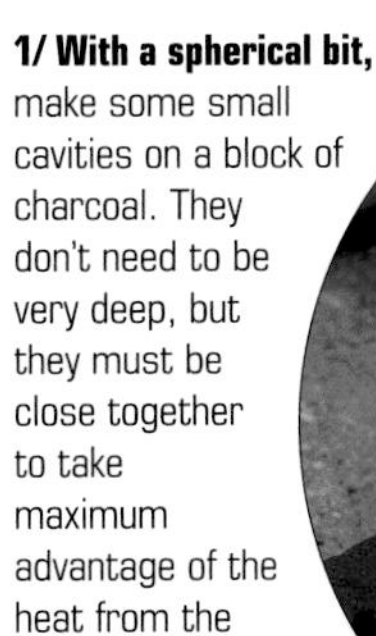

1/ Use an old pickle solution that contains plenty of copper as a result of successive cleanings. A sheet of copper is put into the pickle and heated up.

2/ The clean granules are put onto this sheet and then a welding rod or a piece of steel wire is placed in the solution. This causes an electric bridge in what is a base electrolyte bath. The result is that the copper in solution is deposited onto the granules. The granules will simply need to be covered with organic glue before being applied.

On a Charcoal Block

One quick and efficient way to make spherical granules of an even diameter is to make them directly on a charcoal block. This technique is useful when you don't need to make a large number of granules.

1/ With a spherical bit, make some small cavities on a block of charcoal. They don't need to be very deep, but they must be close together to take maximum advantage of the heat from the torch.

2/ Small rings made from round wire are moistened with liquid flux and put inside each hole.

3/ Starting at one end, the torch is used to melt the rings into granules. Apply a large bushy flame that will melt the rings in groups and produce good, spherical granules. Finally, they are pickled and rinsed with distilled water.

Granulation Paste

The purpose of granulation paste is to add the copper necessary to cause fusion on the surface of the granules. When the temperature is increased in the presence of this copper, the metal is agitated thermally and groups of crystals break apart. At this precise instant, the copper penetrates the structure and creates a tangential alloy with a lower melting point than that of the original metal, giving rise to an interpenetration of the structures and joining the elements. To create this, the paste must contain two essential elements: copper and carbon.

Any copper salt can provide the copper. Good choices are copper hydroxide, copper chloride, copper acetate, and black (CuO) or red (Cu2O) copper oxide. The carbon is provided by the organic glue, which must be a colloid of organic origin that is capable of producing carbon when it burns. You can use gum arabic, tragacanth, fish glue, or gelatin, and even certain white school glues that are entirely organic.

1/ The two essential ingredients in making the paste are copper carbonate and some type of organic glue or colloid. In this instance, we have chosen powdered tragacanth. We also need distilled water and refined borax in the form of liquid soldering solution.

The Old Method

The following granulation paste, which is used for the projects in this book, is the same paste that was used by the ancient cultures of the Aegean and the Mediterranean. The essential ingredients in this paste were malachite and some adhesive of organic origin. The greenest parts of the malachite were chosen and ground to produce a very fine powder that was mixed with the organic adhesive.

In this chapter, we will use copper carbonate and tragacanth. (Malachite is hydrated copper carbonate CuCO3(HO)2, copper, carbon, oxygen, hydrogen, water, and hydrogen oxide.)

Formula

1 part tragacanth	¾ grain (0.05 g)
4 parts carbonate of copper	3 grains weight (0.20 g)
10 parts borax	8 grains weight (0.50 g)
10 parts distilled water	23 grains weight (1.50 g)

2/ First, the tragacanth is dissolved in distilled water until it has the consistency of yogurt, and it is allowed to sit overnight.

3/ Add another part of borax and three parts of distilled water to the tragacanth paste, mix it up, and let it sit.

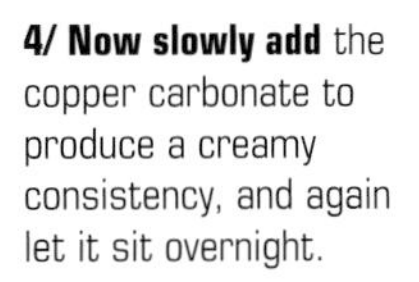

4/ Now slowly add the copper carbonate to produce a creamy consistency, and again let it sit overnight.

The Reaction

Granulation produces a reaction known as reduction, in which the prepared compound is divided into its basic elements. As a result of this, all the oxygen is removed and there is a gain of electrons.

$$Cu_2CO_3(OH)_2 \text{ and tragacanth (resin) } C$$

When it is heated to about 212°F (100°C), the copper carbonate transforms into copper oxide.

$$CuCO_3 \longrightarrow CuO + CO_2$$

When a temperature of 1112°F (600°C) is reached, the resin in the tragacanth turns to carbon and causes the loss of oxygen in the copper oxide. When it reaches 1562°F (850°C), the copper oxide is reduced by this carbon. It turns into copper because of the lack of oxygen. When the carbon burns, it turns into carbonic anhydride.

$$CuO + Cu(OH)_2 + C \longrightarrow 2Cu + CO_2 + H_2O$$

When a temperature of 1616°F (880°C) is reached, the copper melts when it contacts the gold or the silver, and the eutectic bond between the metals is produced. This causes a slight alloy that makes it possible to unite the different metals with one another.

$$Cu(OH)_2.CuCO_3 + C \longrightarrow 2Cu + 2CO_2 + H_2O$$

5/ Mix several times to keep lumps from forming in the paste.

6/ You have to wait a couple of days for the paste to be in perfect working condition. It should have a fine, creamy consistency. To apply the paste, you can slightly dilute it in soldering solution and distilled water.

Application

Granules may be of many different diameters, and they may be applied in a great number of decorative ways, but there are two basic application processes: on a sheet or with one another. Once the granules are soaked in the paste and placed onto the piece, they must be allowed to dry slowly so they stick to the surface and there are no spaces between them. Then, a gentle reducing flame is applied to achieve the glow that indicates that the eutectic bond has been produced.

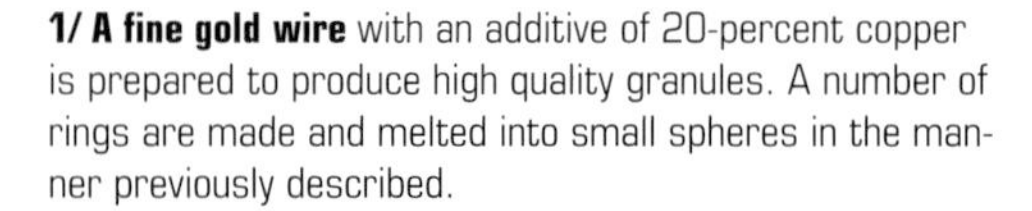

1/ A fine gold wire with an additive of 20-percent copper is prepared to produce high quality granules. A number of rings are made and melted into small spheres in the manner previously described.

2/ Using tweezers or a fine artist's brush, we select some granules of equal size, soak them in paste, and place them onto a fine sheet of mica.

3/ The granules must be completely saturated with the granulation paste. To that end, the paste can be slightly diluted by adding a little soldering solution.

4/ The granules are positioned to form triangular structures of equal size, and they are allowed to dry slowly on the sheet of mica.

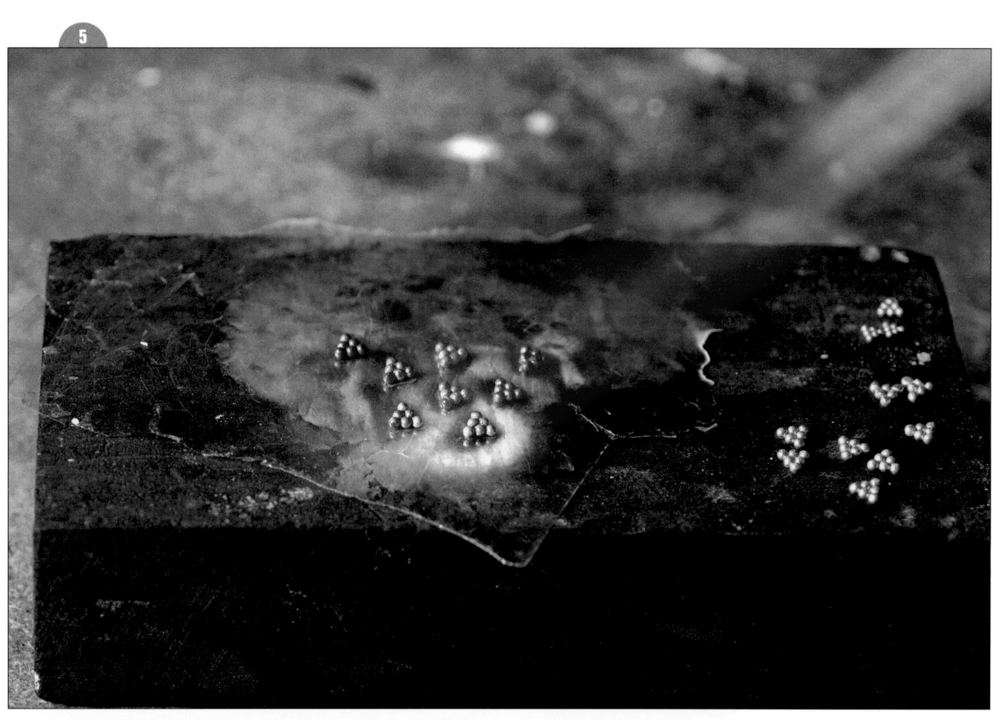

5/ Once they are dry, a reducing flame is applied to each group of granules.

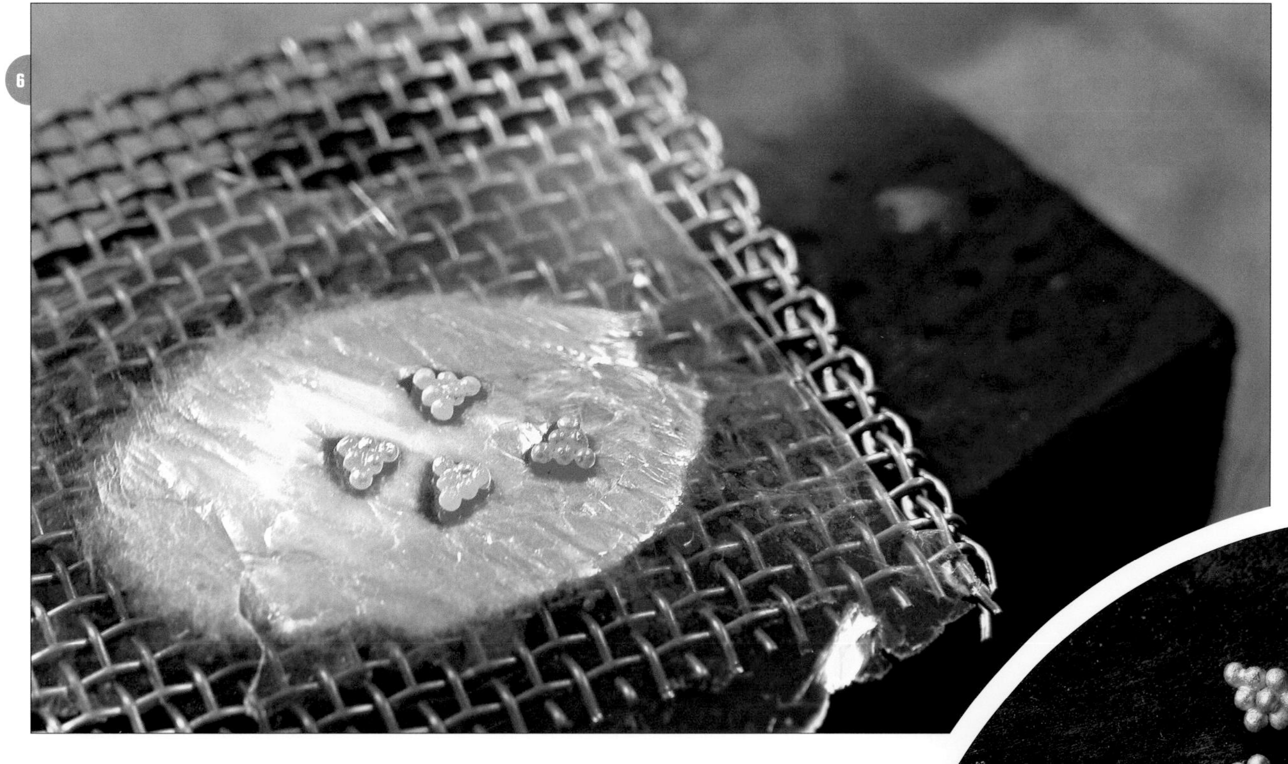

6/ When the temperature is raised, there comes a moment when a rapid, slight glow appears among the granules. This is the precise instant when the eutectic bond is produced. The flame is removed from the granules without delay.

7/ It is possible to create three-dimensional shapes based on basic compositional groups such as triangles or rhombuses.

Making Granulated Earrings

The following earrings are made using yellow 22-karat gold granules. The granules are joined together, and then they are bonded to an 18-karat gold sheet. Solder is used only to attach the post and the rings made of pure gold wire.

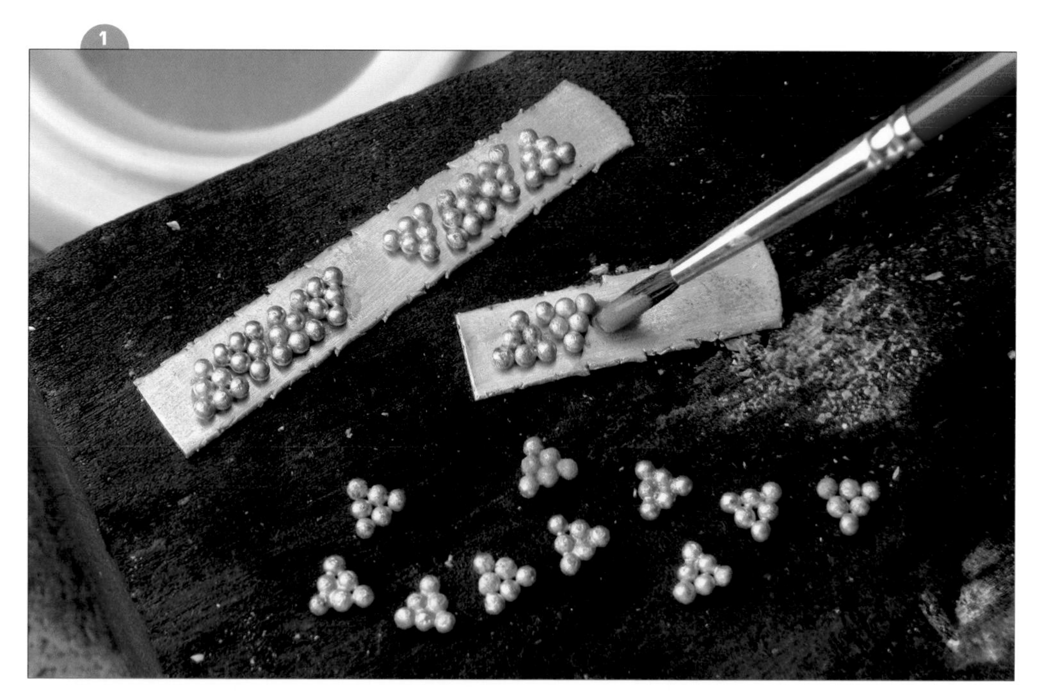

1/ The groups of triangles are applied to the gold sheet with granulation paste.

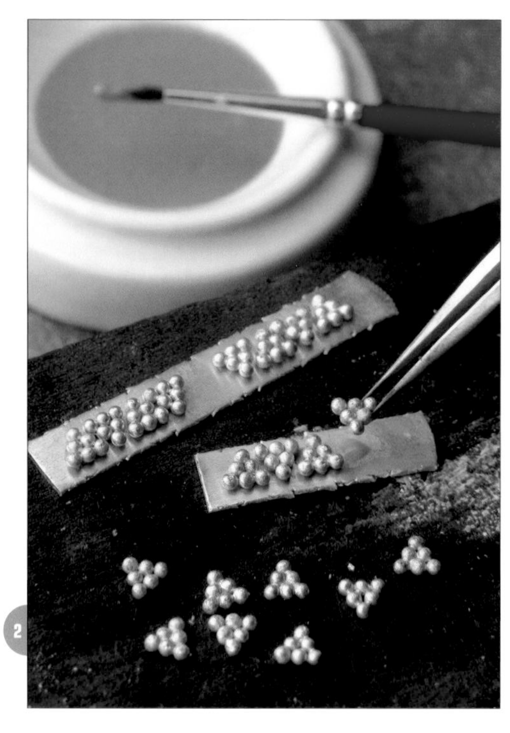

2/ While the paste is still damp, each triangle is put into position and allowed to dry completely.

3/ Heat is applied in the manner previously described, the excess sheet is cut off, and the entire outer perimeter is filed.

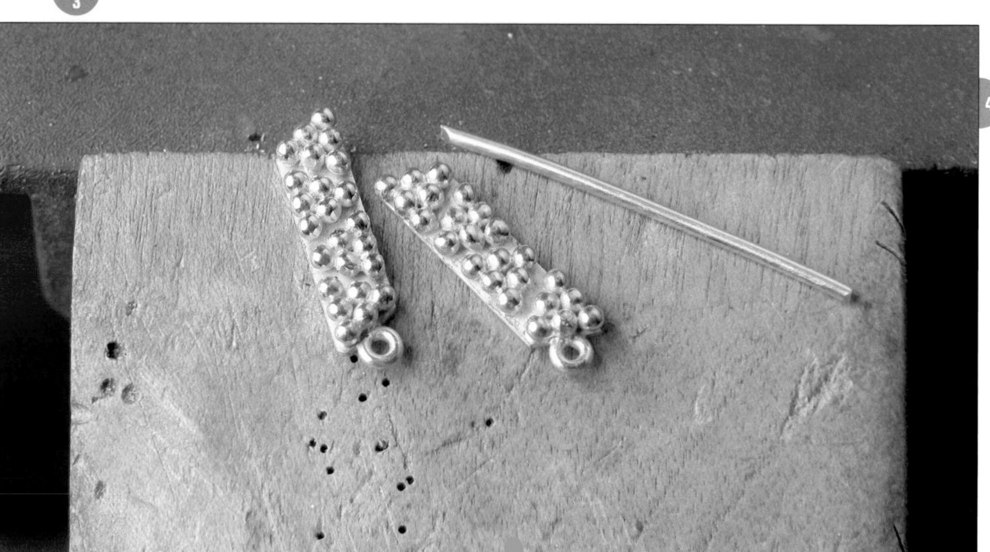

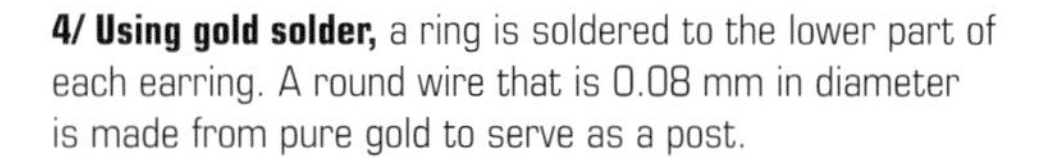

4/ Using gold solder, a ring is soldered to the lower part of each earring. A round wire that is 0.08 mm in diameter is made from pure gold to serve as a post.

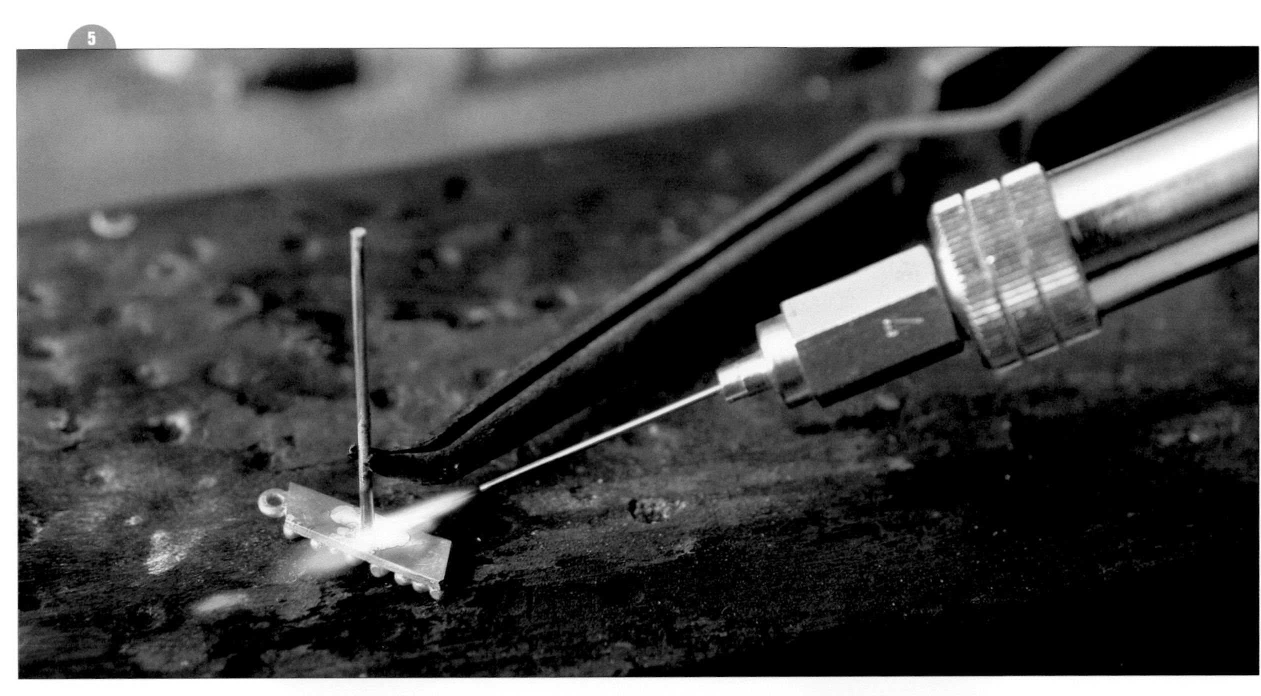

5/ Solder is applied and the post is attached to the back of the earring.

6/ A round wire is also soldered to the center of two rhomboidal structures made up of joined granules.

7/ Two handmade glass beads are added, and a loop is made in the wire with round-nose pliers.

8/ Once the earrings are polished, they are pickled and cleaned to produce the proper coloration for this type of alloy.

Textures and Finishes

With finishes and textures, it is important to be aware of the many possibilities that the metal offers. Aside from the traditional bright polish, there are a multitude of other finishes you can create without using a polishing machine. These range from manual sanding or scraping to sophisticated chemical treatments. Creating a different finish involves searching for uncommonly interesting surfaces. Textures produced through a variety of methods can give a piece personality and make it truly different. This is the goal of the following chapter. Without getting into the most complex finishing procedures, such as galvanic baths, we will focus on some easy processes for producing textures and finishes on the surface of the metal.

Akiko Kurihara. *Inside Animals* rings, 2006. In silver and gold wash (2.5 x 2.5 x 1.5 cm)

Textures

Precious metals such as gold, silver, and copper are softer than nickle silver or steel. When they are put together in a rolling mill, this difference in hardness is used to texture the surface. The pressure the mill exerts causes the design on a hard metal to imprint on the softer metal.

It is also possible to produce textures by striking the surface of metal with hammers that have diverse shapes and profiles and with steel chisels or punches. The metal to be imprinted must first be annealed, and generally the tool must be harder than the metal.

Carmen Amador. Brooch made from silver with elements of gold, coral, quartz crystal, and onyx

Metal elements such as steel mesh, broken jigsaw blades, and steel wires are some of the materials that make a strong impression on silver.

All papers, from a soft paper towel to thick, heavy cardstock to aluminum foil, leave an impression on metal.

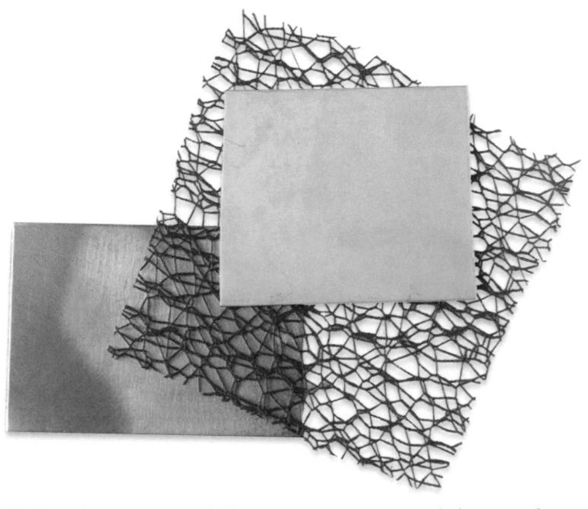

To imprint a texture, it is necessary to anneal the metal first. Keep the texture material from coming in direct contact with the cylinder of the mill by protecting it with a copper sheet.

The rolling mill exerts pressure to imprint the texture on the sheet of silver or gold.

The results of imprinting various materials onto silver.

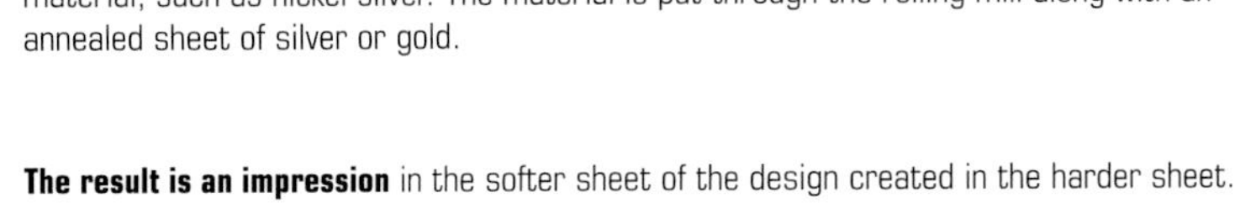

Another imprinting procedure involves cutting out various shapes from a plate of a harder material, such as nickel silver. The material is put through the rolling mill along with an annealed sheet of silver or gold.

The result is an impression in the softer sheet of the design created in the harder sheet.

Finishes

Decide in advance what final finish will be used on a piece of jewelry, from a bright polish to a matte finish to a surface scratched with a large or coarse file. It is possible to create beautiful surfaces by polishing the metal with emery paper. You can also etch the surface of the item in acid to create a white coloration in silver or a yellow one in gold. The same jewelry piece can vary tremendously depending on the ultimate finish it is given. Think about the finish, and try different options before considering it done.

There are various types of diamond tools. For this sample, a coarse diamond abrasive plate was used. It created a very interesting pattern of scratches on the surface of the metal merely with circular movements.

You can create different finishes by using different types of scouring pads on the metal surface.

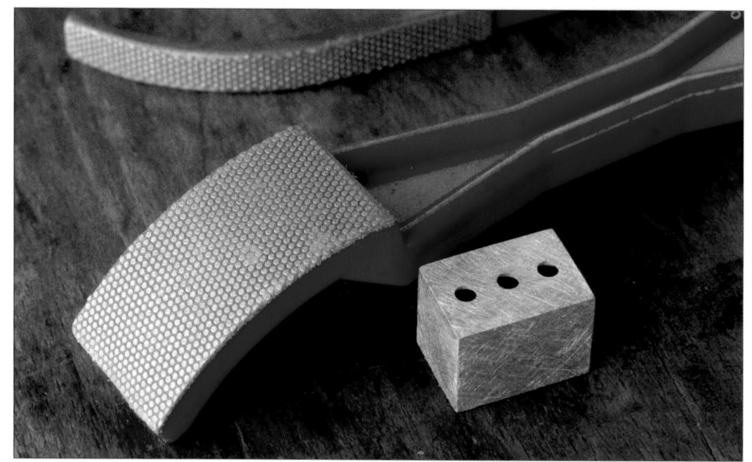

It's not just the coarseness of the abrasive sheet that determines the finish, but also the shape and the direction in which it is applied to the piece. Scratching with a diamond tool increases the shine in comparison to a finish created with emery paper or steel wool.

1/ To heighten the color of gold and give it a hue closer to that of fine gold, the piece is oxidized with the torch, and then put into hydrochloric acid for a few moments.

2/ The piece is removed from the acid and the acid residue is neutralized in sodium bicarbonate. Then, it is rinsed in water.

The results of scouring with a kitchen scrub pad on a red gold and a white gold cube with no subsequent treatment.

The photo shows a yellow gold and a white gold cube with heightened color after being treated with hydrochloric acid.

Pink gold cube textured with paper towel and treated with hydrochloric acid.

Dani Fábregues. Satin silver ring. The metal was scored with a diamond file.

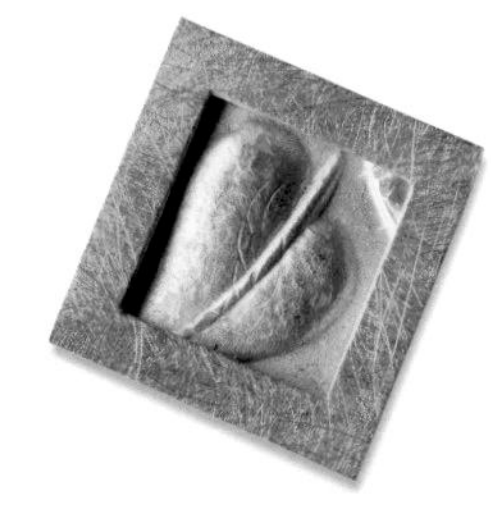

Earrings finished by scoring the metal surface with a diamond cloth.

Stefano Marchetti. Brooch made with a combination of silver and gold.

Blackening

Piece made of sterling silver, and then oxidized with potassium sulfate

Gold does not oxidize naturally, but silver does. It takes on a black color when it comes into contact with environments containing sulfur. When diluted in water and heated, potassium sulfate oxidizes sterling silver and gives it a very interesting black hue. All it takes is submerging the degreased piece in the solution, and then submerging it in boiling water to fix the oxide. Once it is dry, the piece can be sealed with a matte, water-base varnish or a high-quality, hard glossy one. Sometimes it is possible to apply a dark finish on gold or silver by using an electrolyte bath of black rhodium. This gives a special, beautiful, dark quality to the metal.

Patinas

Precious metals, specifically silver and alloys with a high copper content, react in various ways when attacked by atmospheric agents. Patinating metal does not involve applying a surface layer of some product, like painting. It involves an oxidation reaction, a process that may be natural, or as we will see in the following pages, it can be controlled to produce the desired color and texture. Silver naturally oxidizes with the sulfurs in the air and turns black. In the same way, sculptural bronze turns green through the effect of pollution, the acids in animal waste, or contact with rainwater. In the jewelry workshop, it is common to oxidize silver or produce beautiful colorations in bronze alloys by using various chemical products.

1/ Potassium sulfate serves to oxidize the silver black, but it also oxidizes many bronze alloys. The potassium sulfate crystals are dissolved in hot water, and then the resulting liquid is applied with a synthetic brush or through immersion. The bronze and silver come out totally black.

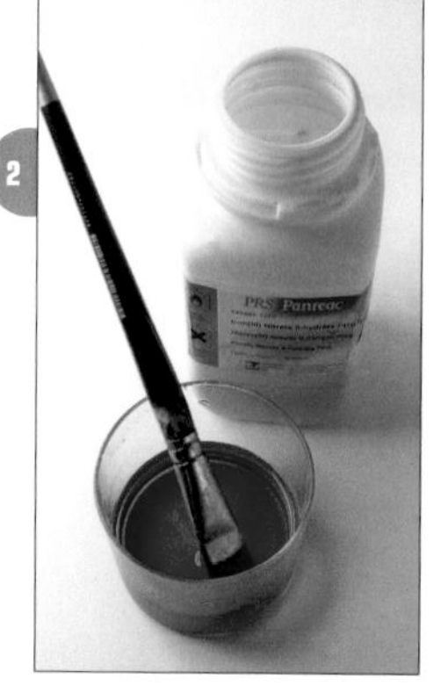

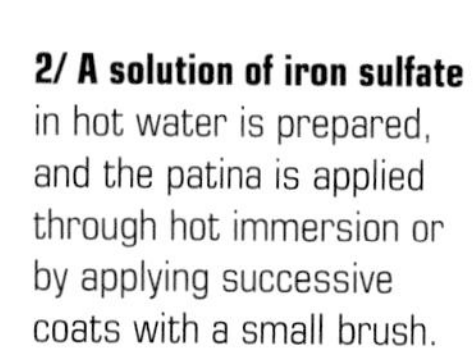

2/ A solution of iron sulfate in hot water is prepared, and the patina is applied through hot immersion or by applying successive coats with a small brush.

3/ The piece is immersed in a hot bath and fixed with a torch or dried in the sun or on a hotplate. If you want to produce different color intensities, the concentration of iron can be changed, as can the immersion time.

4/ Once the successive coats of patina are fixed with heat, the piece takes on an oxidized hue because of the action of the iron sulfate on the potassium sulfate. This change also occurs in patinas made with other products such as sal ammoniac and copper nitrate.

5/ The application of titanium oxide in different concentrations produces a whitish patina on the iron nitrate base. After fixing it with heat, it takes on different cream shades that can vary with new applications to produce the desired shade.

Colors

Light Greens. Successive coats of sal ammoniac diluted with distilled water are applied with a sprayer. The solution is allowed to dry slowly between layers, in the sun if possible. Once the green hue is achieved, it is possible to create ochre and yellowish hues if the pieces are dipped in iron sulfate diluted with water in different concentrations.

Bluish Greens. Copper nitrate is made by dissolving pure copper in nitric acid. Once the acid is saturated, successive coats are applied by spraying or with a brush. They are fixed with heat to achieve the desired intensity. It is also possible to create ochre hues by dipping the pieces in iron sulfate.

Another effective solution involves mixing 3½ ounces (100 g) of copper nitrate in about 2½ cubic inches (40 cm^3) of 70 percent nitric acid in 1 liter of distilled water. The mixture is applied directly to a clean metal piece and heated with the torch or on the hotplate.

Purples. 7 ounces (200 g) of copper nitrate are put into 1 liter of water and brought to a boil. The piece is immersed for 20 minutes and allowed to dry with the same vapor or else it is slightly heated.

Blacks. 77 grains weight (5 g) of potassium permanganate, 1¾ ounces (50 g) of copper sulfate, and 77 grains weight (5 g) of iron sulfate per liter of water are prepared. Everything is brought to a boil, and the piece is immersed for 20 minutes. If you want a glossier, intense black, make a solution of 154 grains weight (10 g) of iron nitrate per liter of water, immerse it, and fix it with heat on a hotplate. Successive coats can be applied to darken the color. The final surface is waxed.

Orange-brown. Prepare a solution of 0.88 ounce (25 g) of copper sulfate and 3 cm^3 of ammonia in 1 liter of water. Apply it through immersion in a boiling bath.

Orange-red. Dissolve 1¾ ounces (50 g) of copper sulfate, 77 grains weight (5 g) of iron sulfate, 77 grains weight (5 g) of zinc sulfate, and 0.88 ounce (25 g) of potassium permanganate in 1 liter of water. Apply through immersion, boiling in 3-minute intervals, and remove the black coating that forms each time. Then, immerse the piece for 20 minutes until it takes on the desired color. The final surface is waxed.

6/ The result after brushing successive layers of iron sulfate onto the heated piece to give it shades of ochre.

7/ The iron sulfate is applied to pieces previously patinated by using sal ammoniac, copper nitrate, or iron nitrate to produce a yellowish base. It can be prepared in different concentrations and applied through immersion or with a brush. The piece is allowed to dry slowly in the sun or on a slightly heated plate.

8/ Bracelets patinated through successive applications and drying with sal ammoniac and water. Once the light green shade was achieved, they were immersed in different concentrations of iron sulfate. Finally, the surface was waxed.

Direct Hot Application

1/ To make sure that the surface of the metal is cleaned properly, it is essential to do a preliminary pickling.
It is also necessary to clean the piece with a wire brush and bicarbonate.

2/ Then the metal is immersed in hot potassium sulfate or in the desired solution for a few minutes. After fixing with heat, the piece will be totally black. Iron nitrate is applied to produce a brown shade similar to iron oxide. Then, titanium oxide dissolved in water is applied with a brush to lighten the patina.

3/ In order to improve the ochre color, different solutions of liquid iron nitrate are applied hot on a hotplate at minimal temperature until the desired shade of ochre is achieved.

4/ Once the desired shade is obtained, the piece is coated with natural wax to improve and protect the patina.

Keum Boo Pendants

The keum boo technique was used as the foundation for making the following pendants. The purpose is to show how to apply the sheet of fine gold, how to produce a texture, and how to finish an object made using keum boo. We will also show the construction and the findings used for attaching the different elements of the structure, plus how to make various gemstone settings and use different soldering procedures.

1/ The first step is to cut out a disk from an 0.8-mm sheet. Then, prepare it to receive the application of the fine gold. The surface is scoured with a piece of fine emery paper and degreased with a stiff brush and bicarbonate.

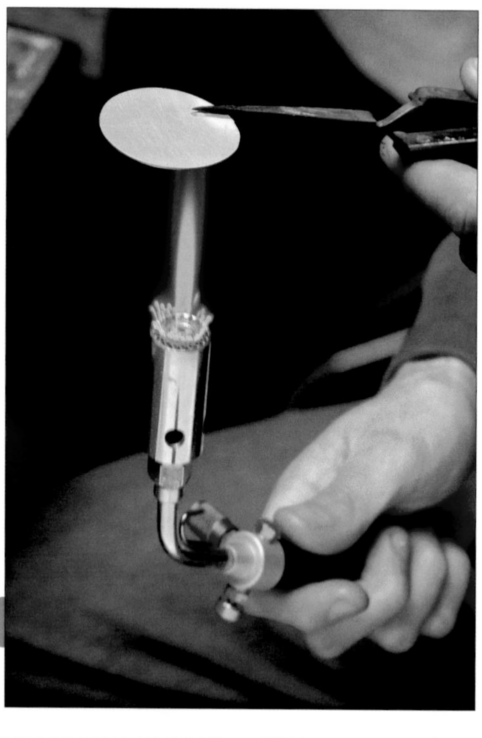

2/ The disk is heated with the torch and dipped in pickle. This depletion-gilding process is done some ten times in succession, producing a surface layer of fine silver that will make it possible to attach the sheet of fine gold.

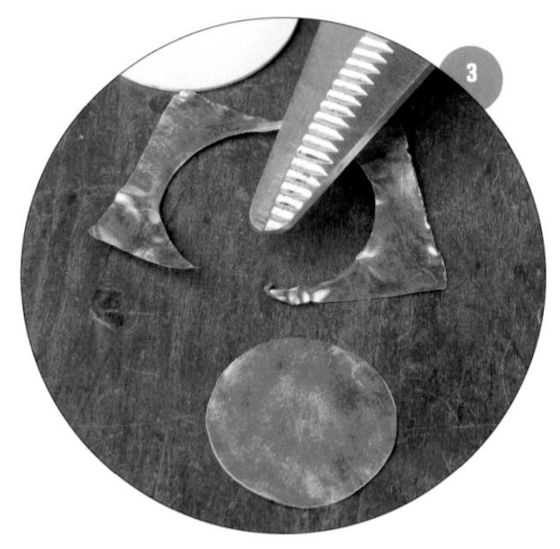

3/ The fine gold is rolled as thin as possible in the rolling mill. It is then rolled between two sheets of copper to produce a thickness of 0.02 mm. A circle that is a smaller diameter than the silver disk is cut out.

4/ Using tweezers and a steel burnishing tool, the fine gold sheet is applied to the sterling silver on the warm burner of a hot plate.

5/ The burnishing tool is moistened in water and used to spread out and burnish the gold on the silver, eliminating any wrinkles in the metal.

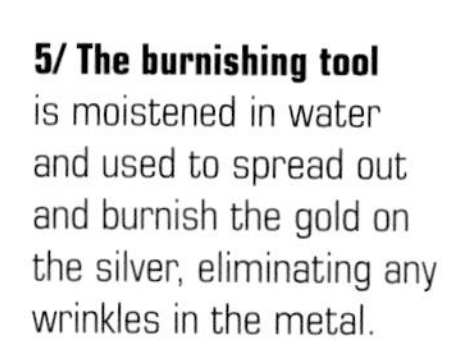

6/ Once the piece has cooled, it is put through the rolling mill with a thick piece of paper towel. The result is a fine texture that unifies the surface and sets the gold into the sterling silver.

7/ A disk cutter is used to cut a circle from the inside of the larger disk.

8/ The disk is worked with a broad steel punch, taking care to put the paper towel between the steel and the silver so the texture is not damaged.

9/ An 0.8-mm-thick sheet is formed into a ring with the same diameter as the interior circle that was cut out of the disk.

10/ The ring is soldered onto an 0.8-mm sheet.

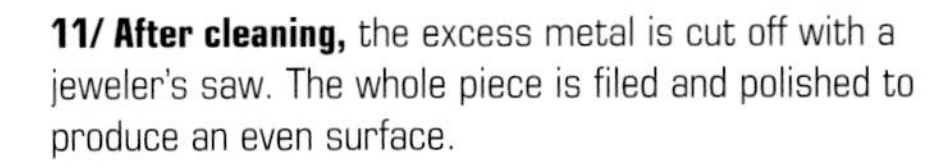

11/ After cleaning, the excess metal is cut off with a jeweler's saw. The whole piece is filed and polished to produce an even surface.

12/ The keum boo disk is fit inside the sterling silver bezel.

13/ Another 0.8-mm-thick silver disk is cut out with a slightly larger diameter than the previous one.

14/ The silver disk is worked lightly in the steel dapping block.

15/ After fitting them together, the silver bezel and the silver dome are soldered.

16/ To make the bail for a chain or steel cable, two pieces of silver tubing are cut, each with a thickness of about 0.7 mm.

17/ The ends of the tubes are filed perfectly smooth. A cylindrical groove is made in the end of one tube.

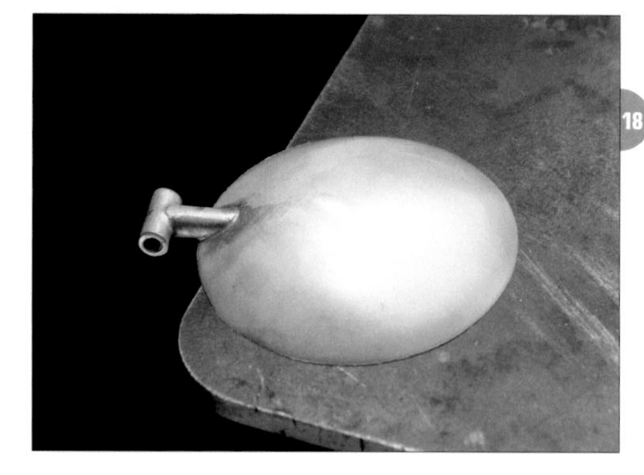

18/ The tubes are soldered with hard solder onto the back of the pendant. The piece is cleaned.

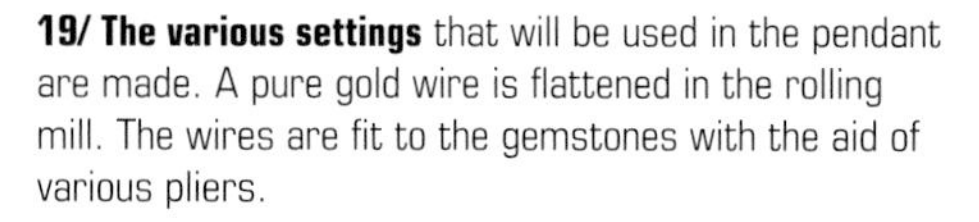

19/ The various settings that will be used in the pendant are made. A pure gold wire is flattened in the rolling mill. The wires are fit to the gemstones with the aid of various pliers.

20/ The setting is soldered onto a fine sheet, and the excess is cut off. A 2-mm silver wire is soldered to the bottom of the setting. The bezel is decorated by cutting out small triangles with a triangular file.

21/ The different gem settings that will be used in the composition are made in the same way.

22/ The gold disk is put inside the bezel, and the holes are drilled for attaching the various settings. The hole for the gems that will be screwed in place must be slightly smaller than the diameter of the steel tap that will be used to cut the threads.

23/ Some of the settings for the gems will be screwed to the back, locking the two structures that make up the pendant. The silver wire is put into a thread-cutting die like the one shown, and the steel die is turned. It cuts a thread into the wire.

24/ To cut the thread in the wall of the hole drilled previously, use a set of threading taps with the same thread pitch as the die used to make the threaded wire.

25/ Three taps are used in order to cut the thread progressively.

26/ Following the same process, the other holes are made using a rotary tool and a bit of the appropriate size.

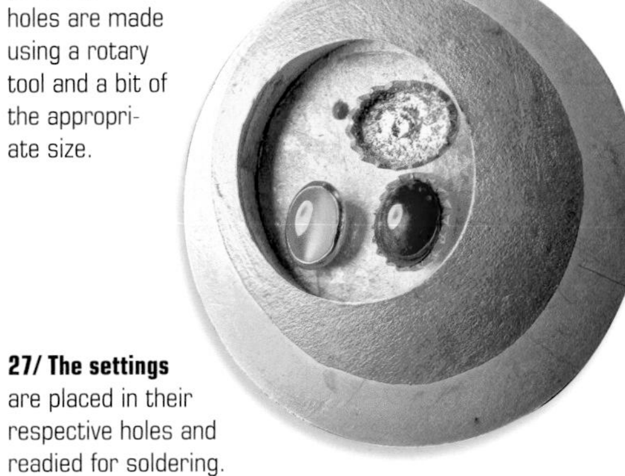

27/ The settings are placed in their respective holes and readied for soldering.

28/ To protect the gems from the heat during the soldering process, a special cooling paste is applied.

29/ Solder is applied to the back of the pendant, and the torch is used to solder the ends of the wires in each hole.

30/ A little cyanoacrylate glue is used to fasten the two parts of the pendant, and the last hole is drilled through both of them. Then, a series of taps is used to produce a concentric thread in the two parts of the piece.

31/ Finally, the back of the pendant is oxidized, and the entire piece is varnished.

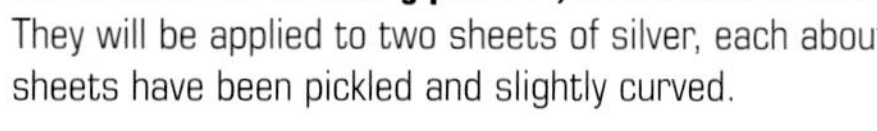

32/ To make the following pendant, two sheets of fine gold have been prepared. They will be applied to two sheets of silver, each about 0.06 mm thick. The silver sheets have been pickled and slightly curved.

33/ Thin gold sheets are heated and burnished on top of the prepared sterling silver sheets.

34/ A craft knife is used to cut out the open work and to trim the outer edge of the fine gold sheet.

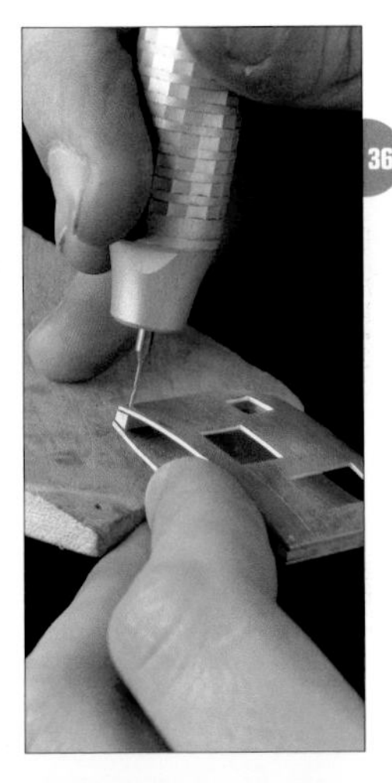

35/ The two sheets are fitted to two square silver rods that will serve as a support and slightly separate the front and back of the pendant.

36/ Holes are drilled near each corner so watchmaker's screws can connect the entire piece.

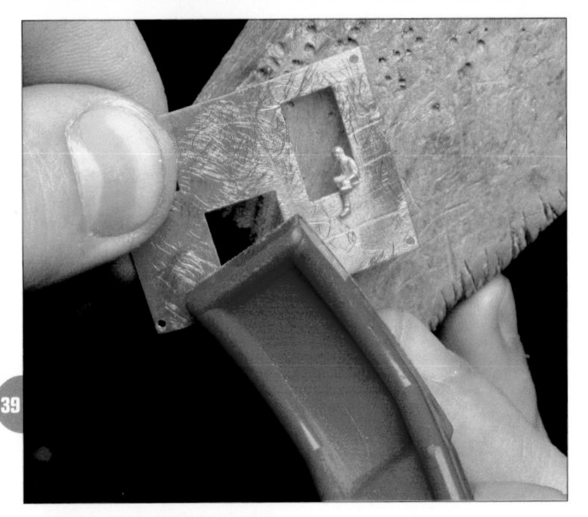

37/ A small silver tube has been soldered to one end of each square rod. The tubes will act as a bail for the steel cable.

39/ To provide the final finish, the surface is textured with a diamond brush.

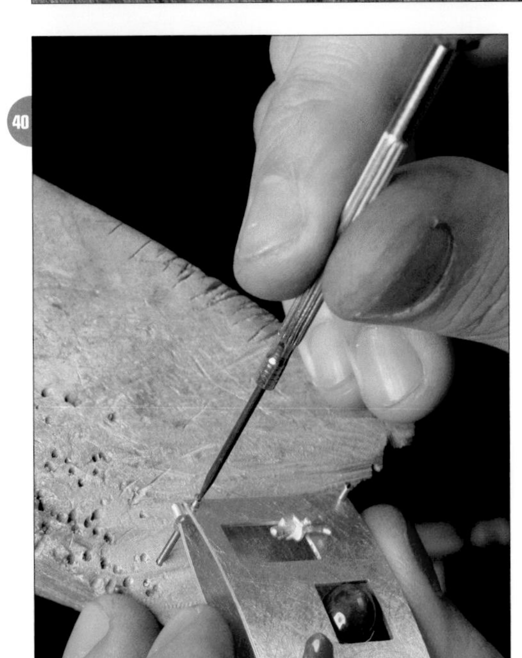

38/ Two bezel settings are made for the gemstones and soldered directly onto the rear keum boo sheet. A small figurine is also soldered to the front sheet of the brooch.

40/ Finally, the watchmaker's screws are turned into the ends of the pendant.

41/ Pendant made by Carles Codina

Granulated Brooch and Ring (Tensi Solsona)

The project by Tensi Solsona contributes something new to the field of granulation. The process allows the artist to make new and interesting jewelry from commercially manufactured and recycled silver items. Instead of causing a reduction reaction to produce copper on the surface, fine silver is added by means of silver chloride previously mixed with the piece. This fine silver makes it possible to bond various sterling silver items without solder, and it produces the right consistency to give the piece solidity. A second paste is also used, which in contrast to the previous one, adds copper through copper sulfate, thereby allowing a more precise way of making this project.

The copper sulfate and sodium carbonate are mixed dry, and the borax is added in powdered form along with the precise amount of water. The mixture is allowed to sit for an hour and is stirred occasionally so that the salts dissolve perfectly.

The mixture causes a reaction in the form of a slight bubbling. When this is completed, silver chloride is added and the two preparations are mixed together. Then, the charcoal is added and stirred to produce a creamy consistency. The resulting mixture is set aside for another hour and is stirred occasionally to produce a perfectly homogenous paste.

1/ The various chemical products used in this procedure must be weighed precisely.

The sodium carbonate, copper sulfate, and borax are mixed dry, and then water is added. The mixture is stirred and set aside for the reaction to take place. The silver chloride is prepared separately by the method described below, as is the charcoal.

2/ Once the mixture has reacted, it becomes an intense blue color. The silver chloride is added and mixed well. Finally, the carbon is incorporated. If the paste is too thick, additional distilled water may be added.

MATERIALS

Powdered sodium carbonate	77 grains weight (5 g)
Powdered borax	61.7 grains weight (4 g)
Powdered copper sulfate	115.7 grains weight (7.5 g)
Silver chloride	15.4 grains weight (1 g)
Charcoal	15.4 grains weight (1 g)
Distilled water	0.60 fluid ounce (18 ml)

Preparing the Silver Chloride

IMPORTANT It is very simple to prepare silver chloride in your own workshop, yet it requires strict safety measures. You must wear acid-proof gloves, an adequate respirator mask, and safety goggles, and you must do this procedure in a well-ventilated area.

1/ Take small silver remnants or filings, and put them into a container. Add nitric acid and water in equal parts. The silver will react with the acid and dissolve.

2/ Once the reaction is complete, common salt is added to the solution by slowly sprinkling it on the upper part of the container. This will cause the precipitation of the silver toward the bottom of the container in the form of silver chloride. Sprinkle in the salt necessary to complete the precipitation of the chloride.

Making the Brooch

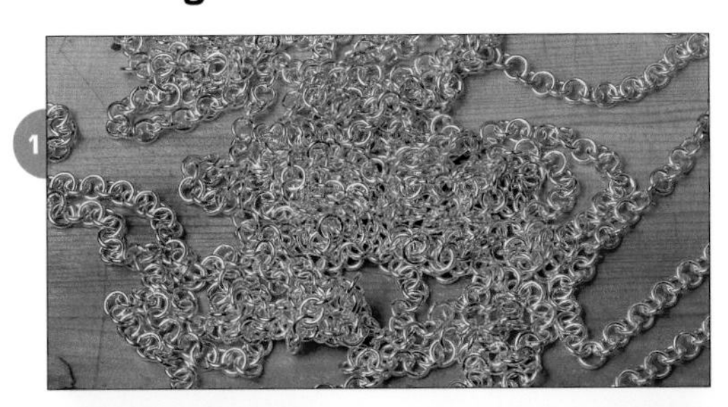

1/ In this project, Tensi Solsona uses a commercial chain made of thick, round rings to make a silver brooch.

2/ The first step is to determine the design and calculate the amount of chain necessary for the project.

3/ To remove all the oxide from the chain, it is annealed and then cleaned repeatedly.

4/ It is essential to clean the silver correctly, produce a surface layer of fine silver, and eliminate all copper from the surface. It must remain free from oxidation and grease to facilitate the bond among all the links of the chain.

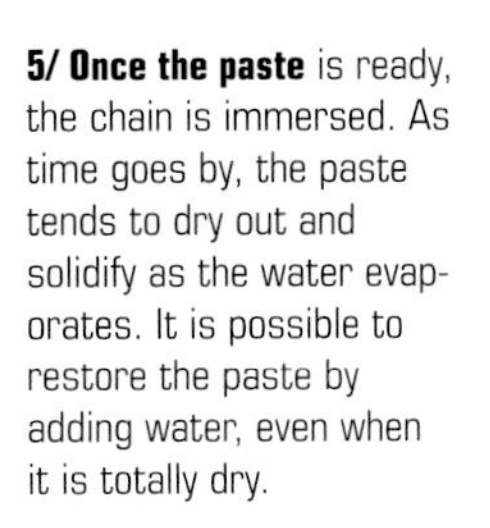

5/ Once the paste is ready, the chain is immersed. As time goes by, the paste tends to dry out and solidify as the water evaporates. It is possible to restore the paste by adding water, even when it is totally dry.

6/ The entire chain is gathered together on a refractory brick to produce the desired shape of the brooch and allowed to dry.

7/ A powerful gas welder is used to heat the entire structure to a red heat. At this instant, the silver contributed by the paste uniformly and strongly joins all the links of the brooch that are in contact with one another.

8/ After applying the heat, the resulting block has remnants of copper slag that must be removed in acid.

9/ To eliminate the slag and slightly thin the silver, a 15-percent nitric acid and water solution is prepared, and the piece is put into the solution.

10/ The piece is heated again and put into the nitric acid. Sometimes it is necessary to replace the acid solution with a fresh one.

11/ The metal is left in the acid solution until the thickness of the rings is reduced to a pleasing size. Then, the metal is rinsed in water and allowed to dry.

12/ If the procedure leaves any rings loose, it can be repeated as many times as necessary. The piece will have to be put back into the acid, new paste will have to be applied to the area, and the part that needs repair will have to be heated with the torch.

Tensi Solsona. Brooch.

13/ The completed metal element.

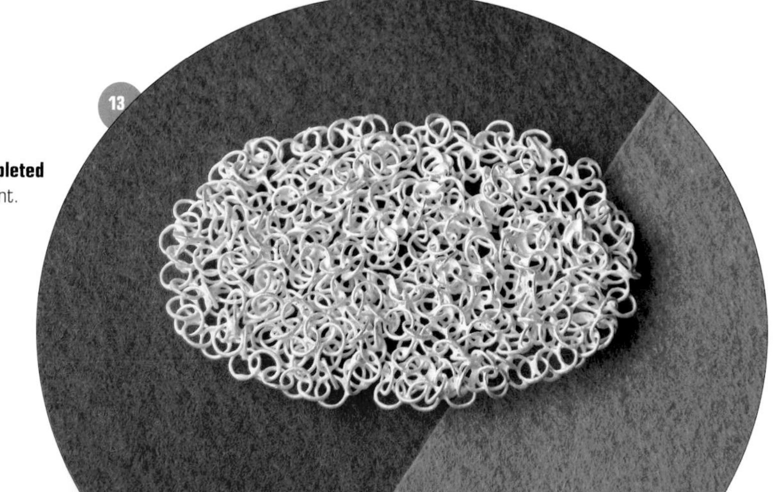

Copper Paste

The second project by Tensi Solsona uses a new paste formulation. Instead of adding silver in the form of chloride, copper is added in the form of copper sulfate. The difference from the previous paste is that it allows more precise work, since copper produces tougher, stronger bonds than silver. The application procedure is similar to the preceding one, and different jewelry can be made in the same way from pieces of recycled silver.

To make the paste, the copper sulfate is mixed with the borax, and the precise amount of distilled water is added with a syringe. The combination is stirred to produce a homogenous mass, which is set aside for an hour and stirred occasionally. The charcoal is added, everything is mixed together, and it is set aside for another hour, again stirring occasionally.

1/ The copper sulfate is mixed with the powdered borax, and distilled water is added to form a homogenous mixture. Then, the charcoal is added.

1

MATERIALS

Powdered copper sulfate	½ ounce (15 g)
Powdered borax	¼ ounce (7.5 g)
Powdered charcoal	31 grains weight (2 g)
Distilled water	0.7 fluid ounce (21 ml)

2

2/ Once the paste is ready, the various chains, wires, or other items to be joined are put into it.

3

3/ The paste must be spread out uniformly on the entire chain before beginning to shape the piece. If the paste is too thick, some distilled water can be added.

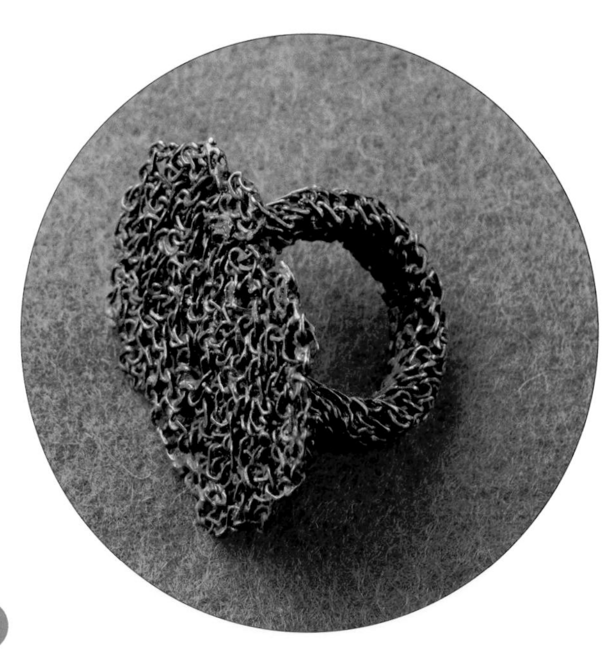

Tensi Solsona. Silver ring made with the addition of copper sulfate.

Mokume Gane Ring (Hans Erwin Leicht)

One of the characteristics that makes the mokume gane technique interesting and unique is not using solder. In the following project, Professor Hans Leicht makes a beautiful ring with this ancient technique. He shows how to prepare a block of mokume gane, and then focuses in particular on the treatment and the manipulation of the metal in order to produce a wedding ring—twisting, forging, and providing the final coloration and finish.

1/ To make the ring, Hans Leicht prepares a total of 13 metal sheets, seven of sterling silver and six of copper. They are cleaned and put into a steel box constructed specifically for this purpose.

2/ The box is put into the furnace, where the sheets bond through diffusion, producing a block like the one in the photo.

3/ To avoid cracks during the forging process, especially in the ends of the block, all edges are filed with a coarse file.

4/ The forging is done at a black heat. (The block is hammered when the temperature decreases and the metal takes on a blackish color.) This procedure must be done repeatedly to produce a 50-percent reduction in size.

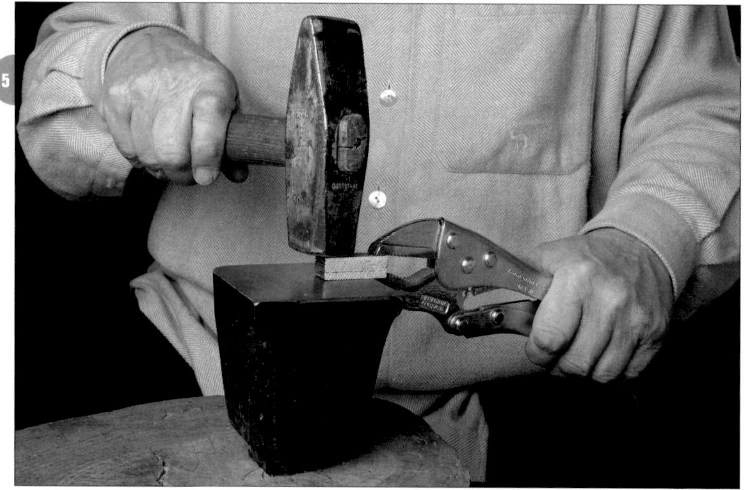

5/ A hammer of at least 18 ounces (500 g) is used for the forging. The block must be hammered evenly and regularly to compact it while reducing its thickness. Proper forging reduces possible cracking and provides better quality and hardness in the metal.

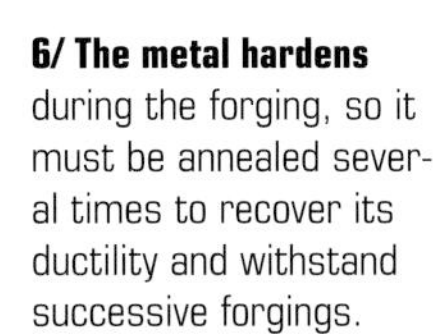

6/ The metal hardens during the forging, so it must be annealed several times to recover its ductility and withstand successive forgings.

7/ Once a reduction of 50 percent is achieved, the block can be put into an ingot mill to square up the ingot and draw it out.

8/ After annealing the block once again, it is pickled and dried. It is secured by one end in a bench vise and twisted longitudinally with a vise grip.

9/ The ridges formed by twisting the ingot are filed, and the metal is drawn or filed to produce a cylinder. Pieces that are 2- to 3-cm long are cut off and drilled in the center so the blade from a jeweler's saw can be inserted.

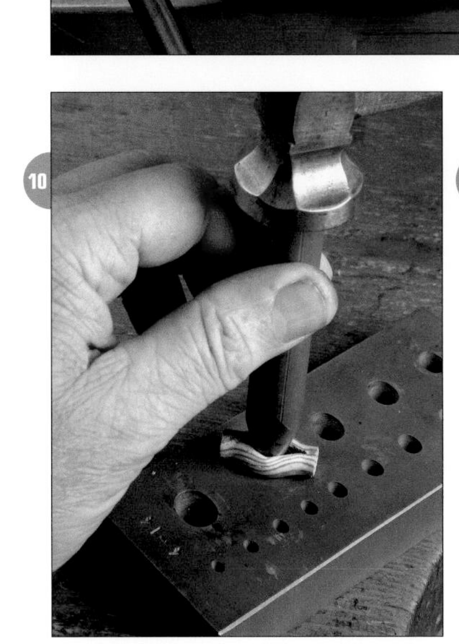

10/ The cylinder is put on a die, and the slit is struck forcefully. This opens up the space in the center so the metal can be slipped onto a ring mandrel.

11/ The exterior of the ring is hammered with a steel hammer to fit the shape of the ring mandrel.

12/ Consistent hammer blows are used so the thickness of the metal is even and the mokume gane design maintains its proper appearance.

13/ The areas that have excess metal are hammered intensively to incorporate the metal into the body of the ring.

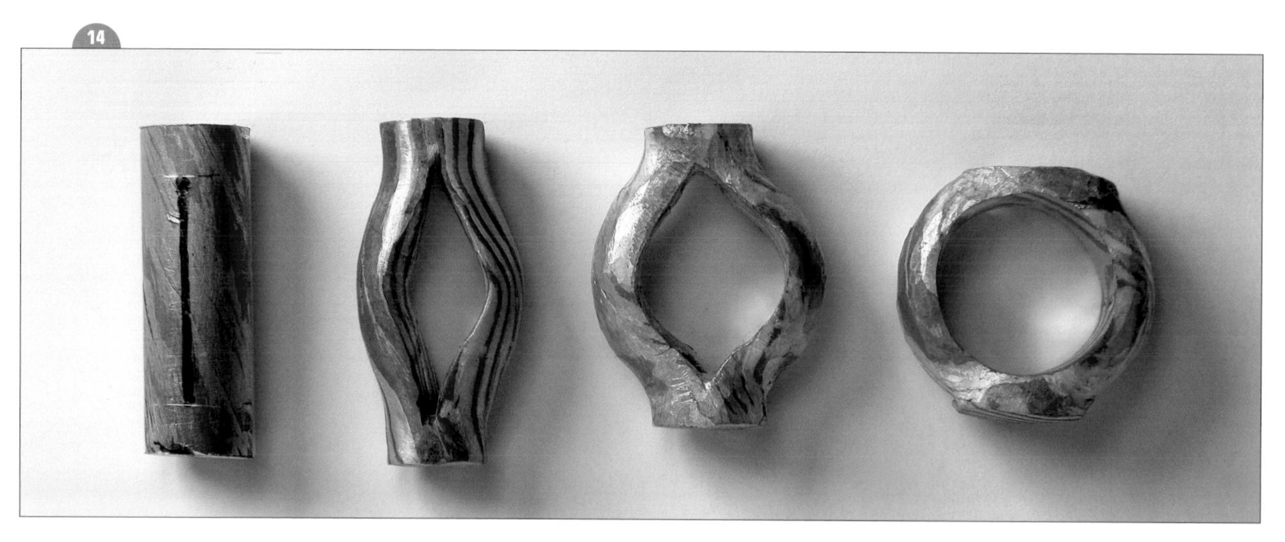

14/ The progressive stages of forging of this ring.

15/ Once the ring is filed and polished, it is ready for the final finish.

16/ A solution consisting of 92.5 grains weight (15 g) of copper sulfate, 15.4 grains weight of salt, and 30.4 fluid ounces (900 ml) of water is prepared. The ingredients are mixed and brought to a boil.

17/ The ring is put into the solution and left there for a half-hour.

18/ The ring can be removed when the patina reaches the desired color. The piece is rinsed in water and given a coat of natural wax.

Hans Erwing Leicht. Mokume gane ring.

Mokume gane rings made by Hans Erwin Leicht using different final colorations

Stamped and Granulated Rings

The following rings combine several of the procedures described in previous sections. First, a texture is created in the rolling mill. Then, decorations are applied with various punches. To conclude, the traditional technique of granulation and the final finish are presented in detail.

This project involves simple construction that requires no complicated tools or complex or expensive procedures. Using techniques and materials similar to those used more than 2000 years ago brings us closer to an anthropological vision of constructing jewelry.

1/ A sheet of gold and a sheet of silver are rolled to a thickness of 0.07 mm. They are fed through the rolling mill with a piece of very absorbent paper towel of fairly substantial weight. It is necessary to use a silver alloy of 0.930 or greater, adding only electrolytic copper to it.

2/ Once the texture is created, the sheet is annealed. It's irregular outline is preserved to make the project interesting.

3/ Select several manual steel punches for stamping various designs on the textured surface of the ring.

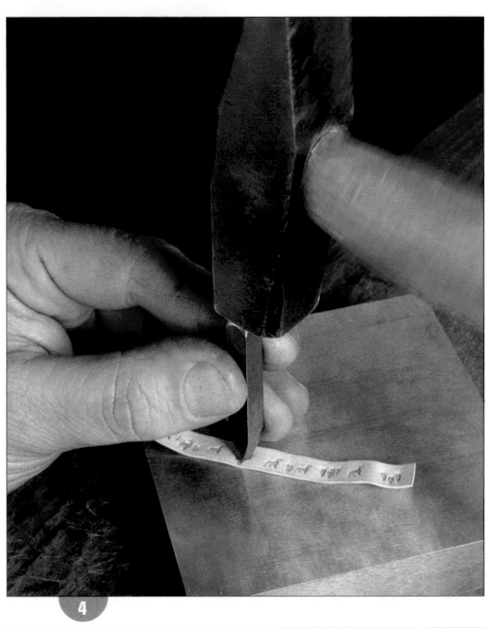

4/ Apply one direct, concise blow from straight above the punch with a medium-weight hammer.

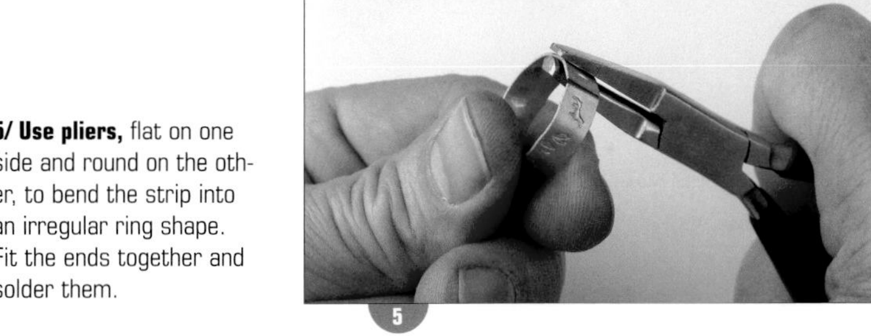

5/ Use pliers, flat on one side and round on the other, to bend the strip into an irregular ring shape. Fit the ends together and solder them.

6/ The rings are soldered with silver or gold solder, depending on the metal used. It helps to bind the ring with steel wire so it doesn't spread apart.

7/ To create a little more thickness on the two edges of the ring and to form a small shiny raised border, they are lightly hammered on a slightly curved swage block, using a hammer with a well polished, spherical head.

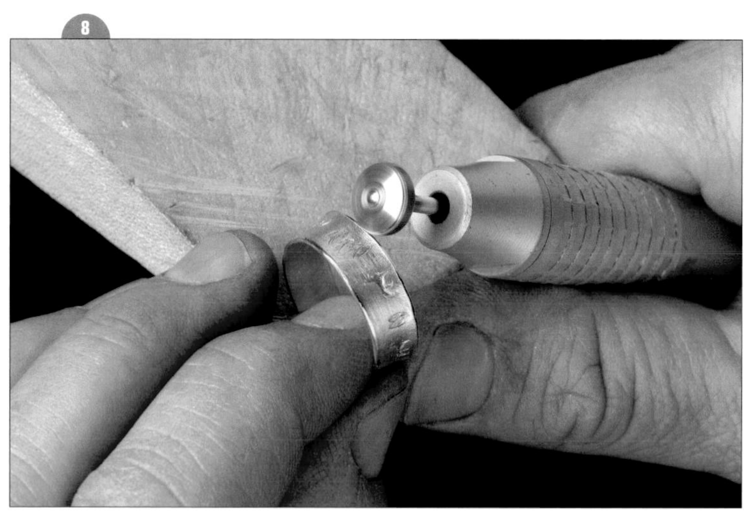

8/ Using a burr, the two metal ridges are burnished in an attempt to raise the edges a little more.

9/ The rings could be left with this finish, but their decoration continues with the application of small granules made from 22-karat gold.

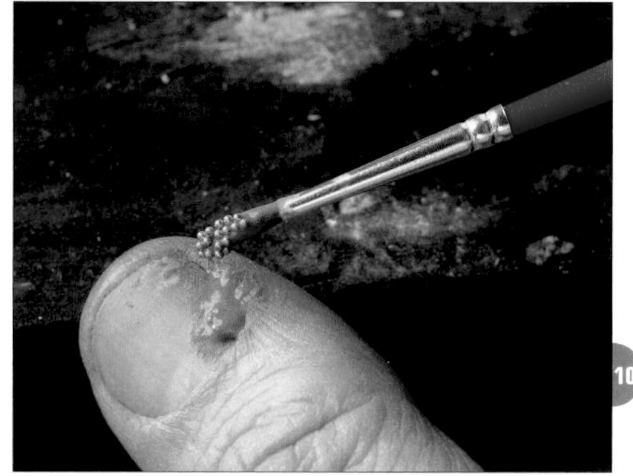

10/ Some granules measuring 0.01 to 0.03 mm are prepared and saturated with granulation paste.

11/ When the granules are very small and an irregular design is desired, it works best to pick them up with a very fine artist's brush.

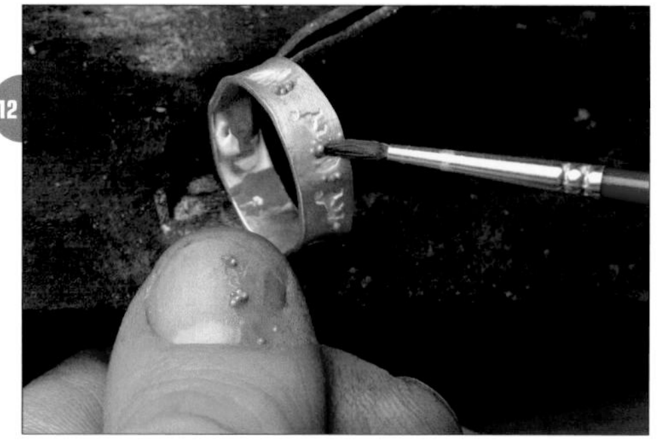

12/ The granules are applied to the metal with the brush, creating good contact between them and the surface of the metal.

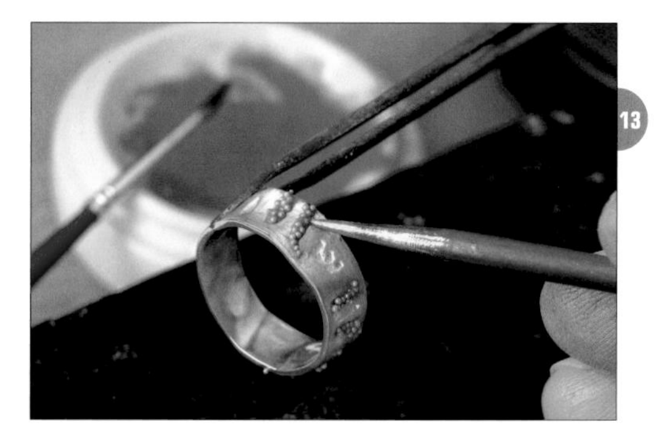

13/ Once they are in place, a steel point is used to create the final pattern.

14/ Very fine tweezers can be used to put the larger granules into place.

15/ A torch is used to apply heat, taking care to use a reducing flame that surrounds the entire ring.

16/ The intensity of the flame is maintained until the granules glow, at which time the torch is quickly removed.

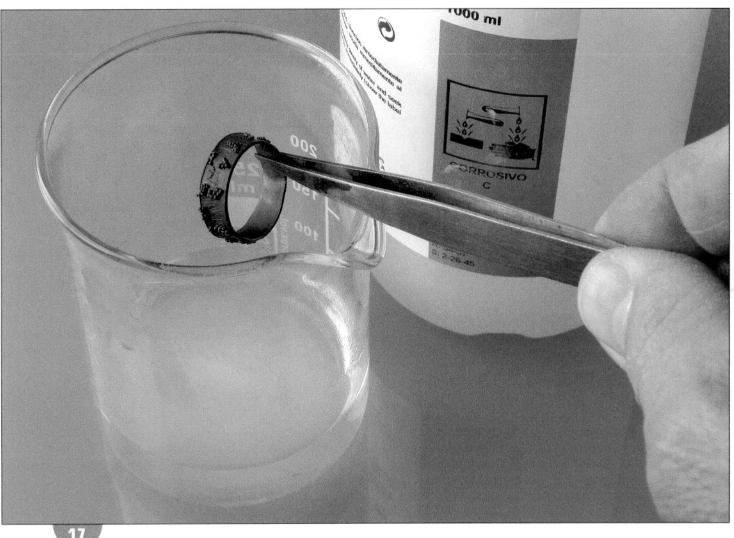

17/ Once all the clusters of granules are bonded with the body of the ring, the metal is heated again with the torch so it will be uniformly oxidized.

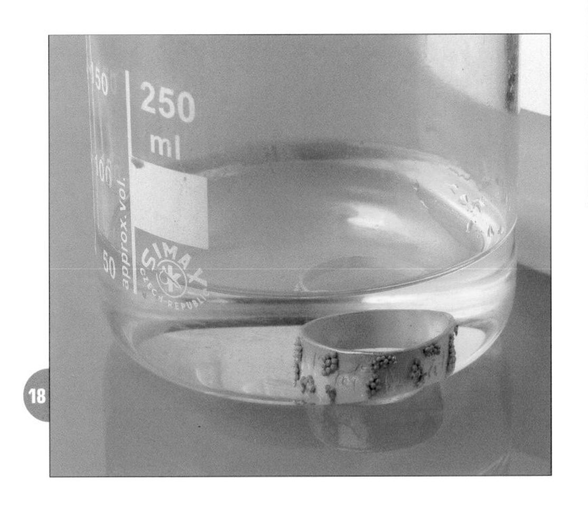

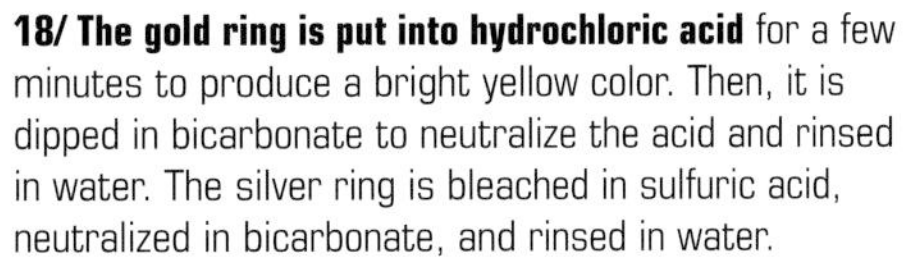

18/ The gold ring is put into hydrochloric acid for a few minutes to produce a bright yellow color. Then, it is dipped in bicarbonate to neutralize the acid and rinsed in water. The silver ring is bleached in sulfuric acid, neutralized in bicarbonate, and rinsed in water.

19/ To finish up, the metal edges of the ring are polished.

PART 2
MODELING AND CASTING
Microcasting and Artisanal Procedures

Introduction

Casting metal and turning it into an object is a magical thing. I feel a special attraction to this process for the immediacy of producing the shape, for the way in which the procedure is prepared, and because it is one of the most ancient ways in which humans have created art objects.

In the first part of this section, we present an introduction to modeling and microcasting. These procedures are still the ones most commonly used to produce jewelry. It is important to know these techniques in order to design and model the pieces correctly and subsequently reproduce them in metal.

In the second part of this section, we explain a series of casting techniques that are based on materials such as dirt, clay, or cuttlefish bone. From small pieces of jewelry to larger decorative items, it is possible to cast directly and simply using these methods. All these techniques have the attraction of immediacy and simplicity in molding the metal, and for me, they have an added value that makes them particularly interesting: the historic character of the preparation and casting procedure. For this reason, I have placed greater emphasis on the development of the process itself rather than on the result obtained. I want to highlight the joy of making a closed crucible by hand using refractory clay and constructing a simple casting furnace with very limited resources. The result is a simple and fun way to understand any casting procedure.

Codinaorfebres
Bronze bracelet patinated with potassium sulfate, iron sulfate, and titanium oxide

Carles Codina i Armengol

Microcasting

Microcasting is the most common procedure for reproducing jewelry. The main advantage of microcasting is that it makes it possible to create a small series easily, quickly, and economically. It is common to sculpt pieces of wax or other materials in order to turn them into gold or silver. It is also common to reproduce multiple pieces from the same pattern for a small production line, or come up with new pieces based on existing patterns to create a full collection. In all these cases, the pattern has to go through a procedure known as microcasting.

The method itself is lost-wax casting. It involves producing various wax pieces and putting them together into a coated mold that can withstand high temperatures. Inside this coating, the wax is removed through heat, and molten metal is introduced into the space left by the wax. Many jewelers do not do this procedure in their own workshop. Rather, they entrust the pattern or the wax copies to a casting specialist who casts them in metal. But it is possible to purchase small-size microcasting equipment at a reasonable price, allowing small studios to carry out this technique on their own.

Molds

To reproduce a piece, it is common to start with a properly prepared pattern or with a piece that has already been made. There are many ways to copy a pattern, although this always depends on its characteristics.

In order to produce a microcasting mold, vulcanizing rubber or different types of silicone are commonly used. These elastomers are found in various formats: silicones that polymerize cold, silicones that vulcanize with heat, and as part of this latter group, silicones with various elasticity and hardness that are appropriate for different projects.

Metal patterns prepared for copying in vulcanizing silicone

Molds made from different types of vulcanizing silicone

A Vulcanizing Silicone Mold

Vulcanizing silicone is used to produce copies of metal patterns. It comes in sheets that are about 7 mm thick and are easy to work with the fingers before they are vulcanized. Through heat applied by the hot plates of a small machine known as a vulcanizer and through the pressure exerted by this machine on the frame, the silicone conforms to the pattern and defines it perfectly. About two minutes of vulcanizing time are needed for each millimeter of thickness of the aluminum frame. Once the vulcanizing time is over, the mold is allowed to cool down, and then it is cut open to remove the pattern.

1/ To make a vulcanizing mold, an aluminum frame with plenty of room for the pattern is needed. Also needed are two plates of the same aluminum to serve as covers for the mold, plus various sheets of vulcanizing silicone.

2/ A runner gate or feeder is soldered onto the pattern. The entire structure is inserted into the frame, which has a small mounting hole.

3/ The pattern is removed and an aluminum sheet is placed behind the mold. Various layers of silicone are cut with a craft knife to fill it loosely.

4/ After removing the plastic protecting the silicone, a first layer is pressed inside the mold with a flat tool. A second sheet was pressed onto the first one for this demonstration.

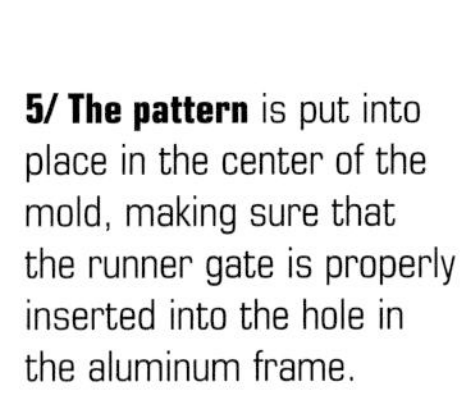

5/ The pattern is put into place in the center of the mold, making sure that the runner gate is properly inserted into the hole in the aluminum frame.

6/ Two more silicone sheets are added. They are compressed by pressing them inside the frame. It is important for a little silicone to stand above the edge of the frame to produce better pressure during vulcanization.

7/ The aluminum cover is put into place, and the mold is put into the vulcanizer. The vulcanizer is programmed to 329°F (165°C) for this type of silicone or to 293°C (145°C) if vulcanizing rubber is used. It's a good idea to consult the tech sheet from the manufacturer. By closing the top of the machine on the mold, heat and pressure are applied.

8/ A craft knife with a triangular point is used to make a straight cut about 7 mm deep in the top of the mold, corresponding to the entry to the runner gate.

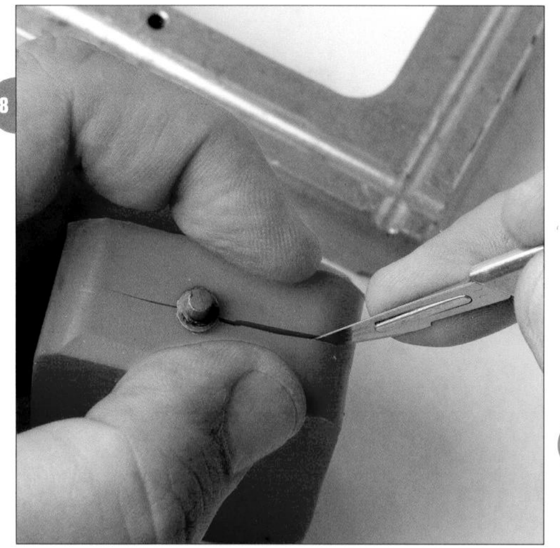

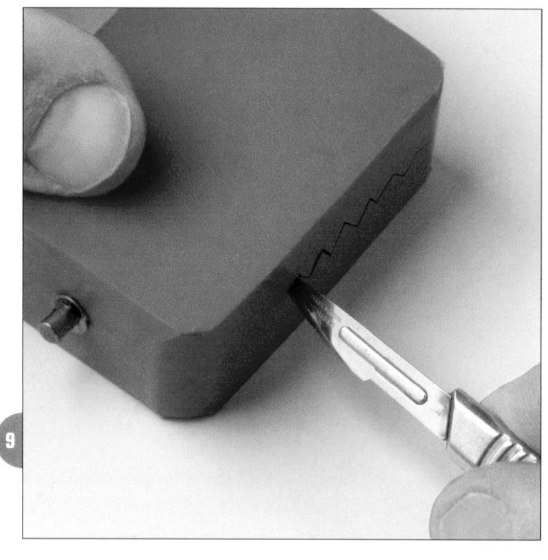

9/ Various triangular cuts are made to the same depth along the entire side of the mold.

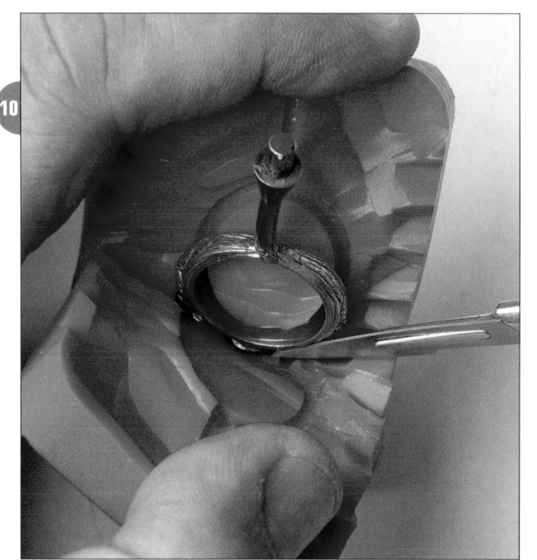

10/ Once the outside cut is complete, the mold is cut from the top down to reach the pattern. The cut extends through the center of the mold so that the pattern can be removed properly and without damaging it.

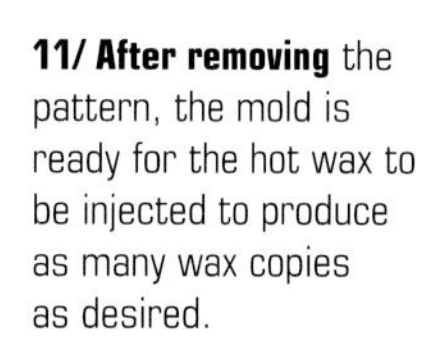

11/ After removing the pattern, the mold is ready for the hot wax to be injected to produce as many wax copies as desired.

A Silicone Polymer Mold

Silicone polymer is used to produce copies of non-metallic patterns, such as waxes, woods, plastic, or other materials that would not survive the temperature or pressure of the vulcanizer. This process involves a chemical reaction and generates neither heat nor pressure. The silicone, which is initially liquid, transforms through a chemical reaction with a specific catalyst into an elastic elastomer that makes it receptive to wax injection.

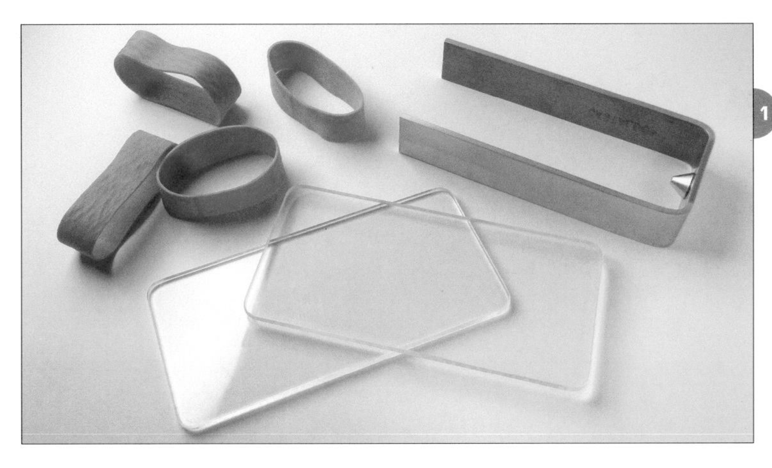

1/ To construct the mold the following are needed: a U-shaped piece of aluminum, a pair of methacrylate sheets that completely cover the pattern, and some rubber bands or a clamp to secure the methacrylate to the shape.

2/ To make a silicone mold from a wax pattern, a small wax rod is affixed to the pattern, and the end is inserted onto a metal cone on the base of the frame.

3/ There are various silicones with different elasticity and mechanical characteristics. For this example, we have chosen a white silicone of medium elasticity, along with its appropriate catalyst.

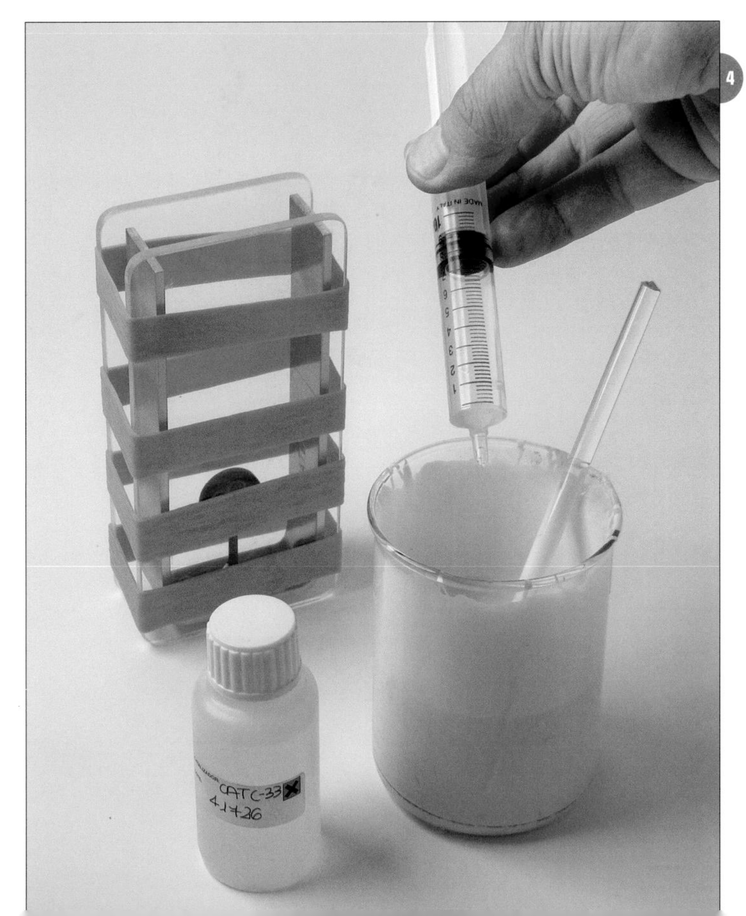

4/ The silicone that will be used in the mold is measured. Once it is weighed, about 5 percent of its weight in catalyst is added. The compounds are mixed without producing bubbles.

5/ Although it is not strictly necessary, it is a good idea to remove the air from the mixture in a vacuum pump.

6/ The mold is filled completely with silicone, taking care not to touch the wax pattern during the pour.

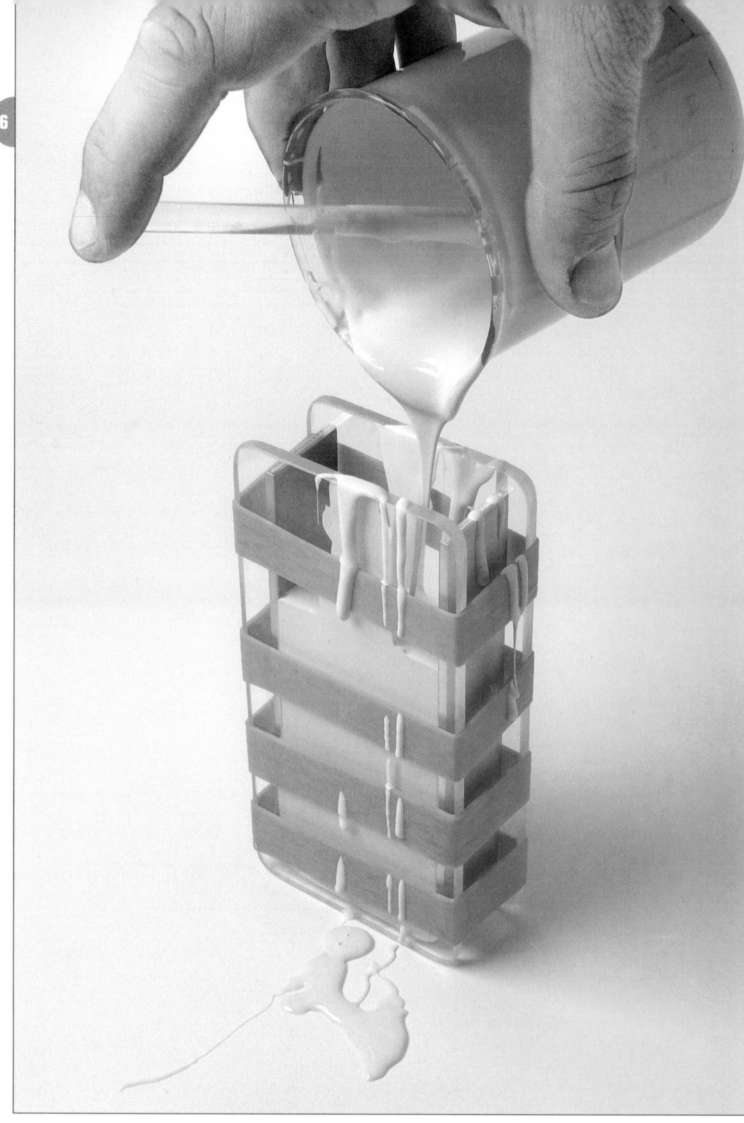

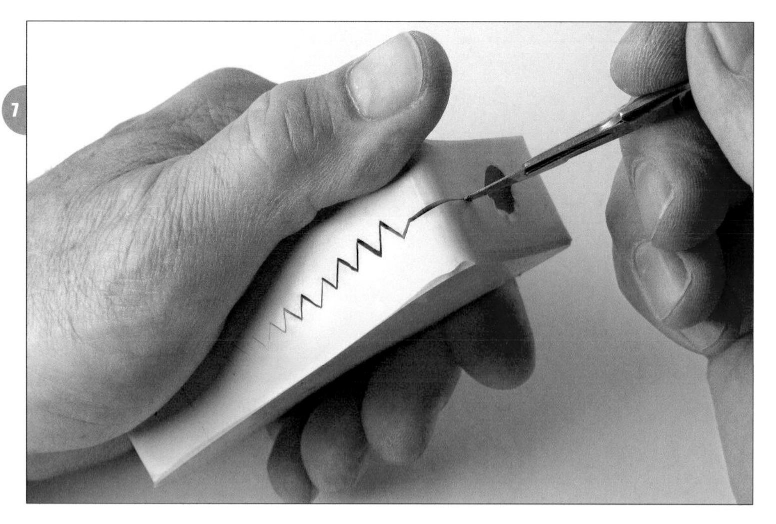

7/ Let the mold sit for a day without moving it. Remove it from the frame once the polymerization process is complete. Cut around the entire outside of the mold.

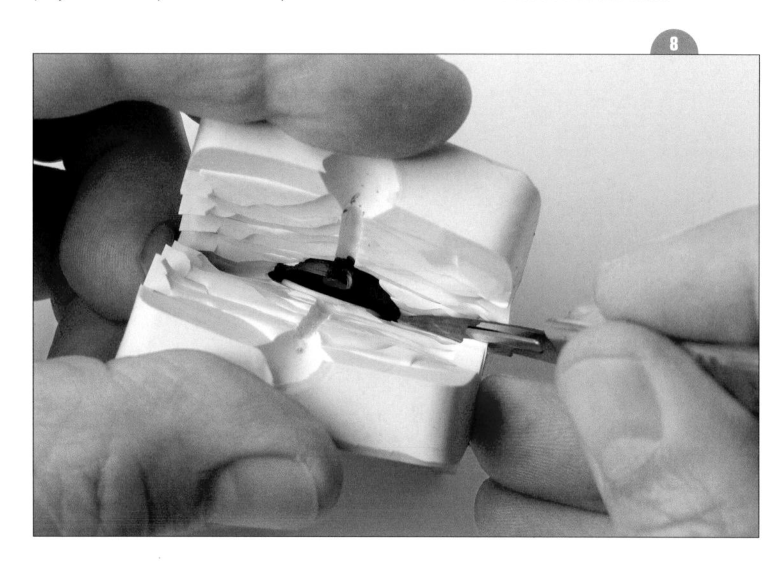

8/ Avoid breaking or cutting the wax pattern during this process.

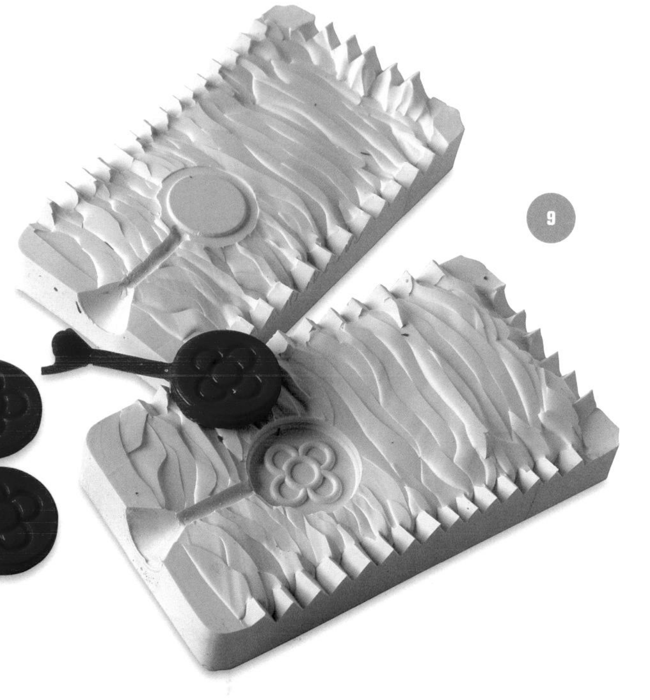

9/ Once the mold is open, you can make as many wax injections as needed.

Injecting Wax into the Mold

Once the pattern is removed from inside the mold, there is a hollow space that perfectly defines the shape of the pattern. Now it is possible to introduce the melted wax to produce as many pieces as desired.

The injector is used to inject the wax. This machine consists of a tank that keeps the melted wax at a specific temperature and a small air compressor that creates air pressure inside the tank.

Different waxes have different characteristics suitable for every type of work. These waxes melt between 149° and 167°F (65° and 75°C), and they leave no residue when burned out in the furnace.

1/ The injection wax is put into the tank, and the compartment is closed. Once the wax is melted at the proper temperature, the air pressure inside the tank is adjusted for injecting the wax into the mold through the valve or piston.

2/ The mouth of the mold is pressed it against the injector piston.

3/ To make a series of copies easily, it's a good idea to sprinkle a fine layer of talcum powder inside the mold.

4/ Clap the two halves of the mold together, and then blow on them to produce a fine, even coating of powder. Then, wax is injected into the mold.

Making the Tree

Once the patterns are injected, the resulting wax positives are attached to a vertical structure known as a casting tree. The assembly is then coated with a ceramic material, known as the refractory coating, which will withstand high temperature. If a centrifuge is used to inject the molten metal, it is preferable to situate the thinnest pieces at the top of the shaft and mount all the wax positives at a 45-degree angle. If casting with a vacuum machine, then attach the patterns perpendicular to the shaft.

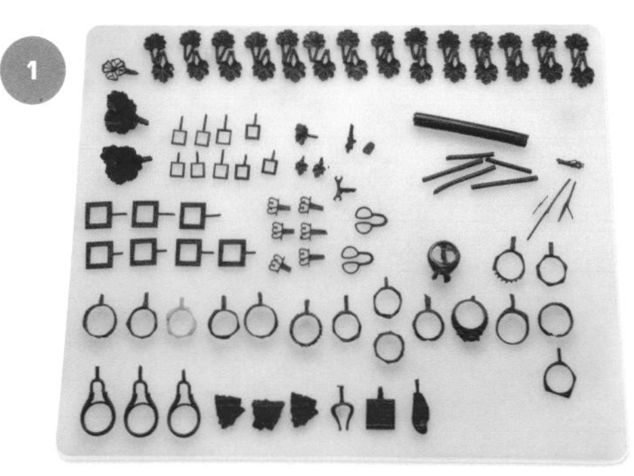

1/ Before preparing a casting tree, it is necessary to check each wax pattern and eliminate or repair the defective ones.

2/ A small rod made from the same wax is attached to each wax pattern to ensure the metal feeds properly during the pour.

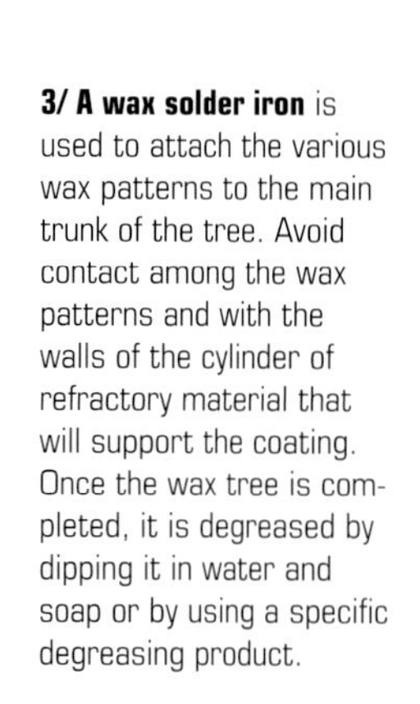

3/ A wax solder iron is used to attach the various wax patterns to the main trunk of the tree. Avoid contact among the wax patterns and with the walls of the cylinder of refractory material that will support the coating. Once the wax tree is completed, it is degreased by dipping it in water and soap or by using a specific degreasing product.

4/ If many pieces are to be cast, it is preferable to group them by size and to use a steel cylinder appropriate for the dimensions of each tree.

Calculating the Metal

In order to calculate the amount of molten metal to be introduced into each cylinder, the wax of each tree is weighed. This weight is multiplied by the density of the metal being cast. An extra 0.5 to 0.7 ounce (15 to 20 g) is added to this calculation to account for the volume of the rubber support base. This is a fairly precise way to calculate the weight of the metal necessary to fill the entire tree. In the case of sterling silver, it is necessary to multiply its approximate density by 10.5. For pure gold, multiply by 15.5. For other materials, it is necessary to multiply by the density of the selected metal or alloy.

5/ It is essential to know the exact weight of the wax before closing up the cylinder. For this purpose, the tree and the rubber base are weighed and the previously calculated weight of the latter is subtracted.

If injecting the metal with a vacuum pump, it is necessary to use a perforated cylinder. Use a smooth cylinder if a centrifuge is being used for the injection.

Coating

Once the tree is finished and placed on the rubber base, the thermal coating is prepared to completely fill the cylinder. The entire assembly will be put into the furnace to burn out the wax tree.

The coating is composed of plaster and some chemical modifiers, essentially silica. The modifier is the most important component for achieving a good cast. It facilitates the elimination of the gases produced during the casting, and it prevents contraction and expansion of the coating. The coating is mixed according to the manufacturer's instructions, in a ratio that varies between 39- and 41-percent water with respect to the quantity of coating powder used. Water, preferably distilled, is used to prepare the coating, at a temperature that never falls below 59°F (15°C). The ideal temperature would be between 68° and 72°F (20° and 22°C). The water is poured in first, and then the coating powder.

An electric mixer is used to stir the mix. This can be a mixer specifically designed for this task or a kitchen mixer with blades that have been filed and slightly bent to keep them from cutting the coating too much. After mixing, the coating is put into the vacuum pump to remove air from inside it.

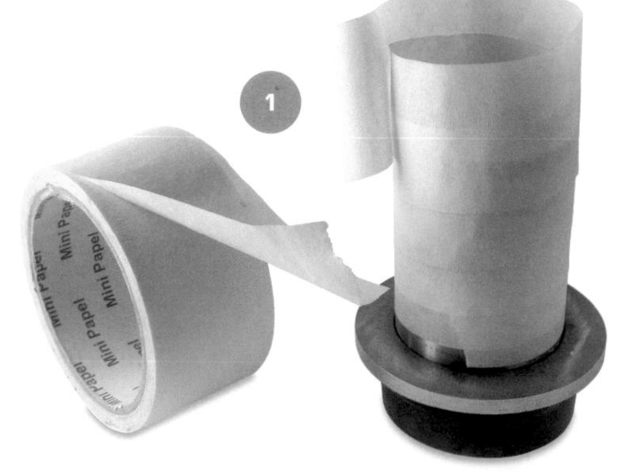

1/ If a vacuum casting machine is used, then use perforated cylinders of refractory steel that can withstand the high temperatures. To keep the liquid coating from coming out of these perforations, they are covered with plastic auto-body tape.

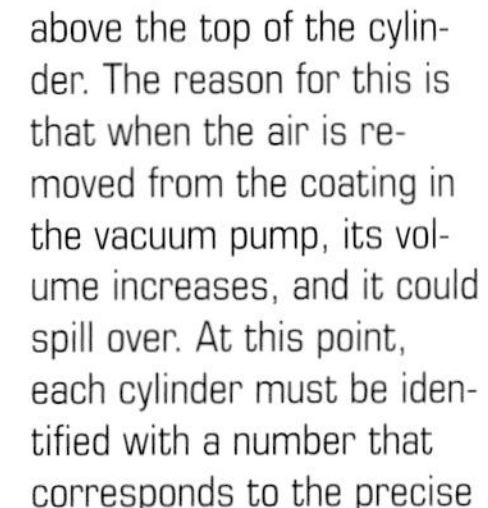

2/ The tape is applied so it sticks up almost 2 cm above the top of the cylinder. The reason for this is that when the air is removed from the coating in the vacuum pump, its volume increases, and it could spill over. At this point, each cylinder must be identified with a number that corresponds to the precise weight of the metal.

3/ The water and powdered coating are mixed in a broad container that will fit inside the bell of the vacuum pump.

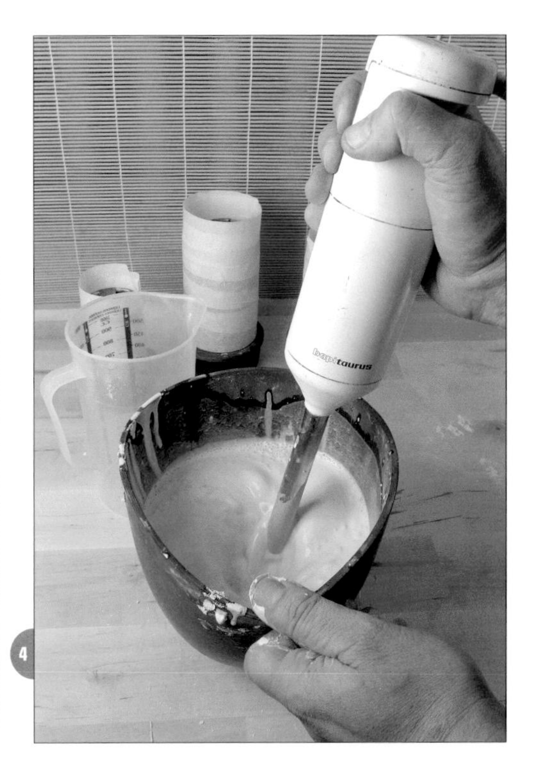

4/ At low speed, the mixture is stirred for a maximum of two minutes. The mixer is continually and energetically moved to produce a homogenous mass of coating.

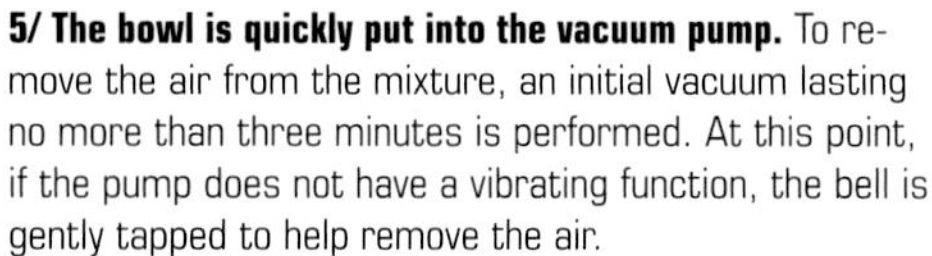

5/ The bowl is quickly put into the vacuum pump. To remove the air from the mixture, an initial vacuum lasting no more than three minutes is performed. At this point, if the pump does not have a vibrating function, the bell is gently tapped to help remove the air.

6/ The pump valve is opened, the coating is taken out, and all the cylinders are filled. It is important to pour the coating into the cylinder carefully to avoid damaging the tree.

7/ Once the cylinders are filled, they are subjected to a second vacuum with less vibration. This eliminates all the air from inside the cylinder.

8/ Once the second vacuum is completed, the cylinders are removed from the pump and set aside for at least one hour. The excess coating is levelled from the top. The cylinders are prepared to be put into the furnace for the burnout.

9/ Left: A smooth refractory steel cylinder prepared for casting in a centrifuge. Right: A perforated cylinder ready for casting in a vacuum casting machine.

How to Determine the Volume of the Coating

To figure out the precise proportion of coating and water needed to fill a casting cylinder, the rubber base is put onto the cylinder and sealed with tape. The cylinder is half filled with water. Then, since it is better not to run out of coating during the fill, 25 percent more water is added to the cylinder. This water is poured into a graduated test tube. The volume obtained in cubic centimeters is multiplied by 100 and divided by the proportion of water recommended by the manufacturer, normally 40 percent. The result is the amount of coating powder needed to prepare a specific cylinder.

Once the volume of each cylinder is known, it is helpful to mark each one with the correct proportions of water and coating.

Burnout

Once the cylinders are prepared, they are put into a furnace so the wax can be burned out. The cylinders are put in the furnace with the open base of the tree facing down so the wax flows vertically toward the floor plate of the furnace. The burnout operation involves raising and stabilizing the furnace temperature several times, and finishing up with a slight temperature decrease, at which time the metal is injected into the mold.

The progressive temperature increase is first designed to remove moisture from the cylinder. Then, as the temperature is increased, the wax liquefies and burns when it reaches 752°F (400°C). When this phase is done, the temperature is raised in order to completely burn up the wax and harden the coating. The final phase involves a slight temperature reduction for injecting or casting the metal inside the cylinder.

The temperature increase must be progressive and must be done in stages, followed by an appropriate stabilization. This must always be adjusted according to the diameter of the cylinder used: the larger the cylinder, the longer the stabilization time. It takes the heat longer to saturate the center of the cylinder where the trunk of the tree is located. If cylinders of different diameters are inside the same furnace, the stabilization time is based on the cylinder with the greatest diameter.

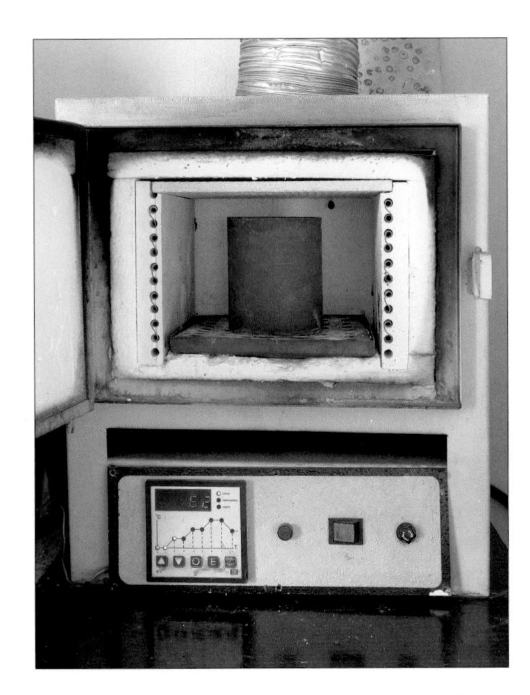

The furnace has a microcomputer for programming the intended heat curve and countdown. All the cylinders are put inside, the heat curve is programmed, and the furnace is turned on.

Temperature Chart

1. During the first hour or hour-and-a-half, increase to 212°F (100°C).
2. Maintain the temperature at 212°F (100°C) for another hour or hour-and-a-half, depending on the diameter of the cylinder.
3. Increase the temperature progressively up to 752°F (400°C) over three hours.
4. Maintain the heat at 752°F (400°C) for a half-hour, based on the largest cylinder diameter.
5. Go straight up to 1382°F (750°C) in a half-hour.
6. Maintain a temperature of 1382°F (750°C) for an hour or an hour-and-a-half, depending on the cylinder.
7. Go back down to the casting temperature in a half hour. Depending on the metal to be cast, this is between 842°F (450°C) and 1112°F (600°C). This final temperature must be maintained for at least an hour for the temperature to penetrate the entire cylinder.

Casting

Every casting process consists of melting a metal and introducing it into a cavity where it will solidify. After eliminating the wax, we merely have to melt the metal and introduce it into the empty space. There are various methods and different machines that facilitate and improve the quality of the casting, from a simple gravity cast to sophisticated and costly machines. In the two methods presented here, we use simple, economical machines within the reach of any small shop—a spring centrifuge and a simple casting machine. Both produce castings of excellent quality.

Vacuum Casting

1/ The vacuum casting machine consists of a vacuum pump that performs two functions. The plastic bell allows the elimination of the air from the coating. In the other compartment, the same pump produces a very strong vacuum in the cylinder and suctions the metal toward the interior of the pump. (Continued on page 78)

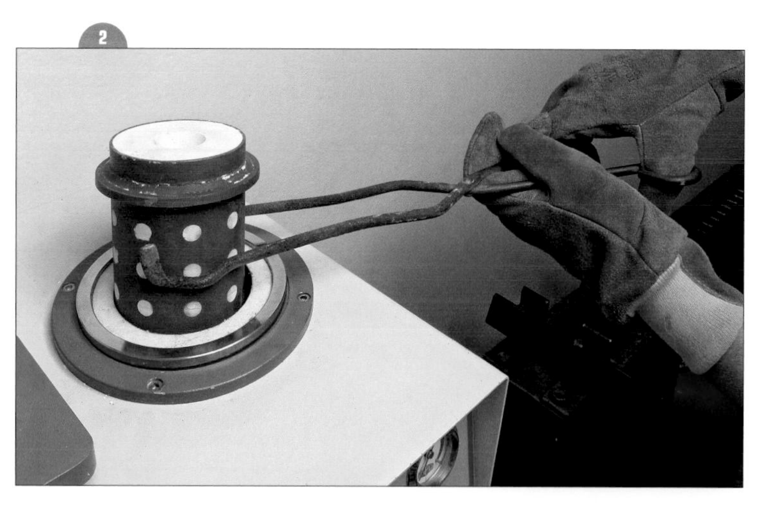

2/ The hot cylinder is taken out of the furnace and quickly put into the vacuum pump, which is charged and ready to create the vacuum.

3/ The vacuum is created inside the casting chamber at the same time that the molten metal is poured into the cylinder.

4/ Once the metal is poured, the cylinder is put into water. The resulting thermal shock breaks the coating and frees up the tree with the parts.

Centrifugal Casting

5/ The centrifuge injects the metal through centrifugal force. This force is released by a motor or a spring and impels the molten metal toward the inside of the cylinder loaded in the crucible.

6/ The metal melts right in the crucible, in this instance with an oxy-propane torch.

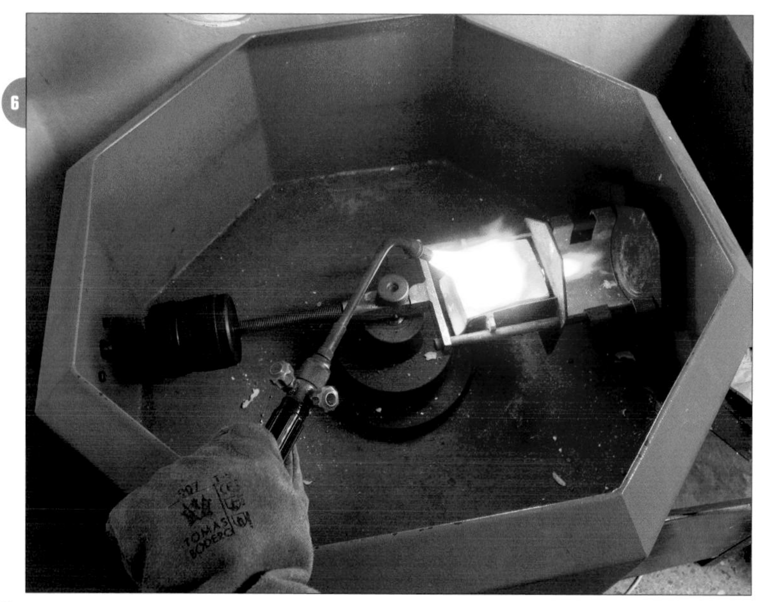

7/ The cylinder is taken out of the furnace, and the base of the tree is fitted to the opening of the crucible.

8/ Once they are fitted together, the arm of the centrifuge is released. It quickly begins to spin at high speed, introducing the metal that was in the crucible into the cylinder through centrifugal force.

9/ After about a minute, the cylinder is taken out of the machine.

10/ When the metal has cooled, the coating can be removed with pressurized sand and water or with a hard brush. The metal tree is put into pickle or acid to remove the resulting oxidation.

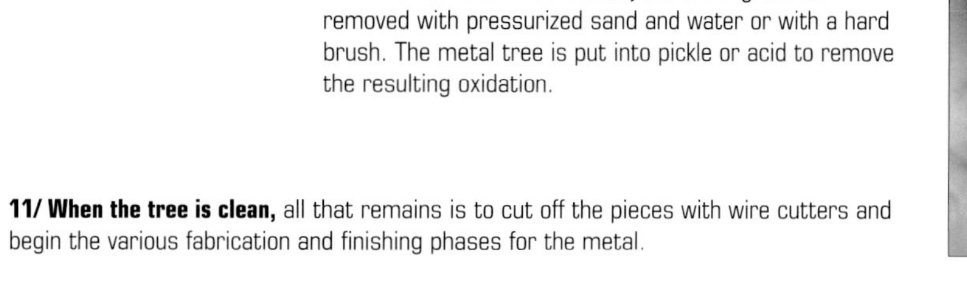

11/ When the tree is clean, all that remains is to cut off the pieces with wire cutters and begin the various fabrication and finishing phases for the metal.

Working with Waxes

Wax is a material that melts quickly at a low temperature and is easy to manipulate. It is used to create a form, and then it is introduced inside a mold or a coating made of a heat-resistant material. Once the wax is removed or burned out with heat, the molten metal is introduced into the space it left, in a process known for centuries as lost-wax casting.

Different types of modeling wax

Modeling Waxes

Modeling waxes have very different characteristics. The most commonly used ones are the rigid or hard waxes that are found in various formats—sticks for making rings or bracelets, cubes of different sizes, and so forth. These waxes can be worked cold, and with the aid of files, cutters, and grinders, they can be sawed and pierced easily. They can also be worked with programmable or manual lathes.

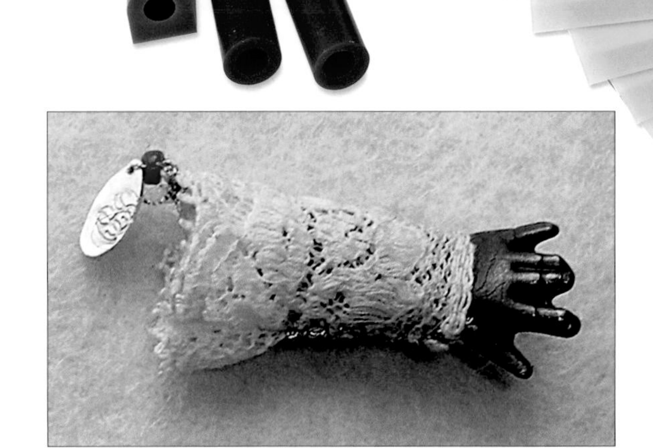

Carolina Hornauer. Brooch made from a casting of a small plastic hand

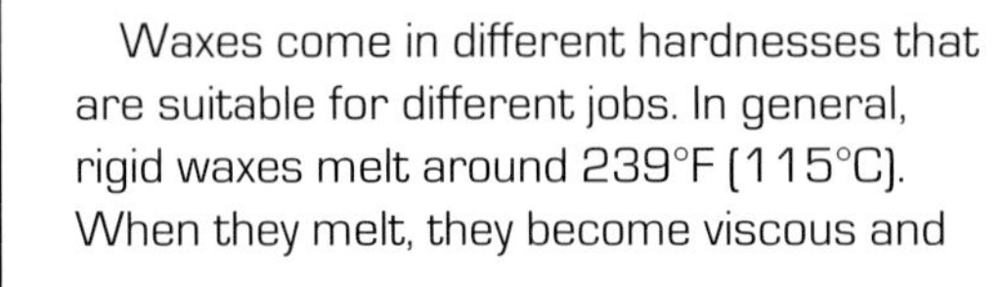

Waxes come in different hardnesses that are suitable for different jobs. In general, rigid waxes melt around 239°F (115°C). When they melt, they become viscous and leave no residue when burned out in the furnace. This characteristic noticeably increases the quality of the result produced in the metal casting.

Elisa Pellacani. Cast pendant made by modeling soft wax directly with the fingers

Soft waxes can be manipulated easily at room temperature. They are easy to adapt and work with the fingers.

A Ring in Rigid Wax

Wax sticks are manufactured in different shapes to use for rings or bracelets. They are very practical for working on a lathe and working by hand.

1/ To model a ring, we begin by adjusting the inner diameter of the wax tube to the desired size. This can be done with a file or other tools, including a hand lathe.

2/ To cut a thickness from the wax tube, use a frame saw with a spiral blade or cut it with a burin mounted on a cut-off lathe.

3/ The wax can be worked in many ways. For this sample, the blue wax has been heated slightly and facets have been cut with a craft knife.

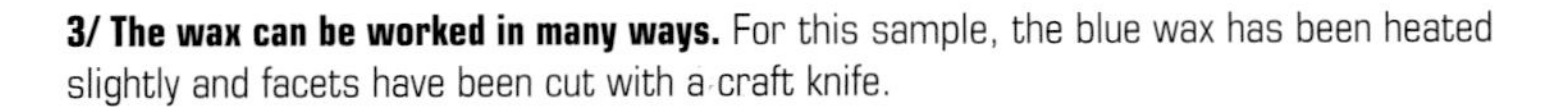

4/ Old files are best for working wax, since the wax fills and clogs the teeth of the file. The facets of this ring were defined and its surface was refined with files.

5/ After filing, very fine emery paper is used to refine the shape, and then the wax is polished by rubbing the surface with a little cotton dipped in gasoline. The wax design is ready for casting in precious metal.

6/ A coating cylinder is prepared with the different wax designs inside, and the casting is done in the desired metal. The surface of the metal is filed and dressed with emery paper before starting the polishing phase.

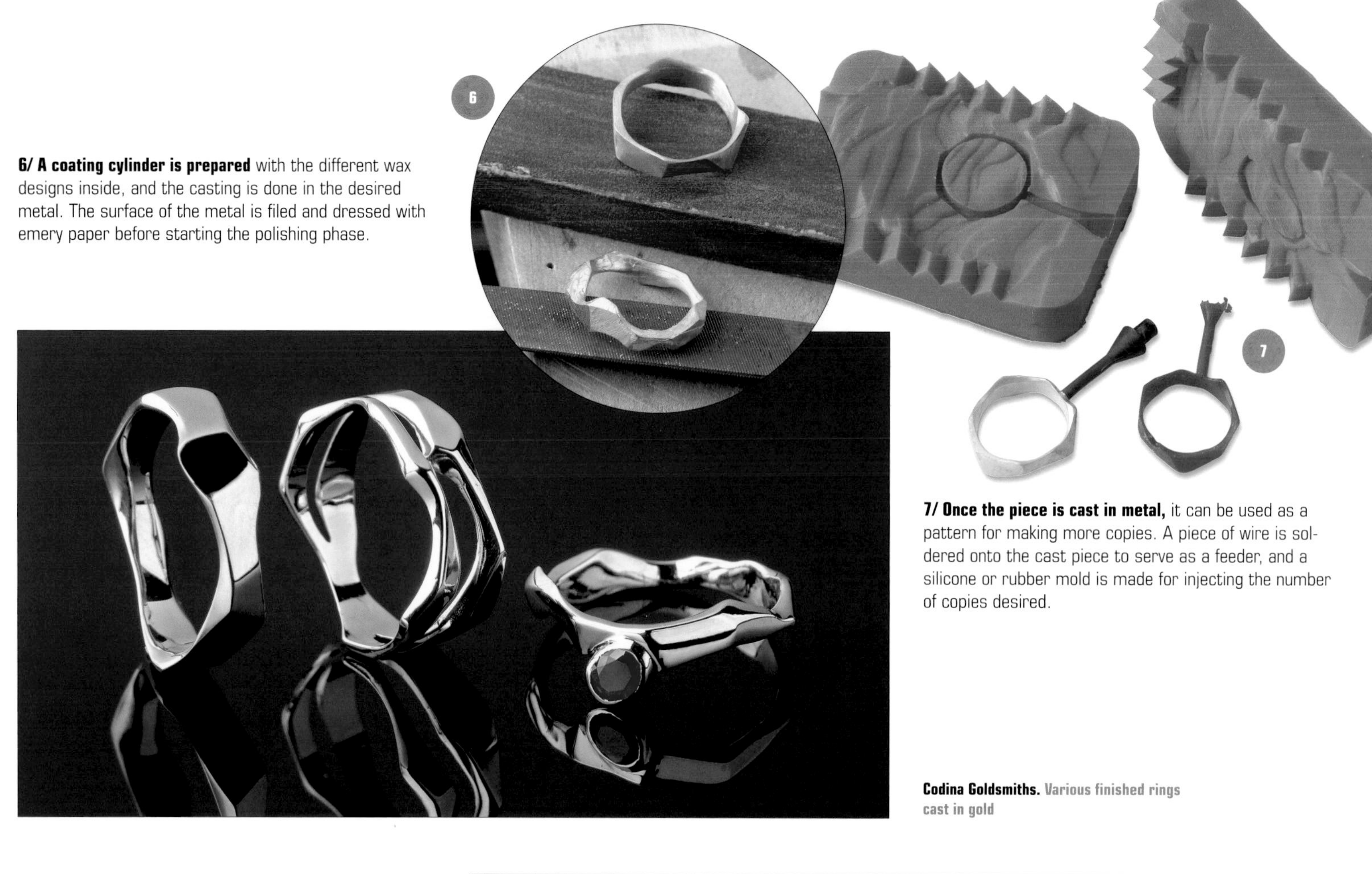

7/ Once the piece is cast in metal, it can be used as a pattern for making more copies. A piece of wire is soldered onto the cast piece to serve as a feeder, and a silicone or rubber mold is made for injecting the number of copies desired.

Codina Goldsmiths. Various finished rings cast in gold

Injection Waxes

These waxes are most commonly put into the tank of the injector that is used with a rubber or silicone mold, but other uses are possible. There are various types of injection wax. Depending on the manufacturer, their melting temperature, viscosity, and hardness vary. These characteristics must be known in order to choose the wax best suited to a particular project. Injection waxes are characterized by their low melting point, between 149° and 167°F (65° and 75°C), their excellent definition, and as with modeling waxes, the fact that they leave no residue after burning out in the furnace. Since they provide excellent definition, they can be used to produce all kinds of shapes and designs. Hot injection wax can even be poured directly onto different surfaces to produce a variety of textures and interesting pieces of jewelry as shown in the following sample.

1/ Many stones can be cast directly, but it is important to choose ones that contain no quartz. Among many others, volcanic rocks, slate, and dolomite are good choices. They all resist thermal shock when they come in contact with molten metal.

2/ A piece of clay is smoothed out, and various stones are sunk halfway into the clay. Textures are produced directly on the clay. A coat of aerosol silicone or a mold release is applied to improve the definition of the wax on the clay surface.

3/ The wax is melted at the lowest possible temperature and poured onto the surface of the clay, where small walls have been erected to contain the wax.

4/ After a few seconds, the wax that is in contact with the surface of the clay is cool. Any hot wax remaining inside the mold is poured off. This way, a fine wax surface is created that will reduce the weight of the piece once it is cast in silver.

5/ The excess clay is removed with water and a brush. If this poses problems, the wax is submerged in vinegar.

6/ The stones are still adhered to the wax. The wax is prepared in a coating cylinder and burned out. The volcanic rock survives the burnout and the metal injection procedure. Then, the silver has to be cut off and polished.

Models for Direct Casting

Natural elements, such as leaves, insects, pieces of wood, and other materials can be cast directly, just like a wax design. In general, this poses no problem beyond the ones related to the shape itself.

Very thin, soft items, such as green leaves and insects, must first be fixed with a lacquer hairspray containing no tars. (When tars burn, they produce imperfections in the casting.) Once the items are fixed, and always based on the characteristics of the item, a layer of melted wax is spread over the surface to increase its thickness. The purpose is to make the item totally rigid, and to produce the thickness necessary for casting. The microcasting procedure requires that the item remain stable inside the liquid coating for the 10 minutes it takes to set. The item must remain still for the proper metal reproduction to be achieved.

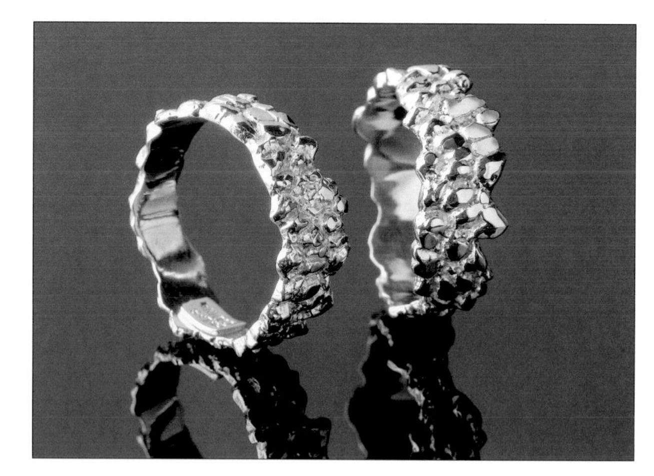

Codina Goldsmiths. Cast gold and white gold rings made using the dried skin of a lichee

Walter Chen. Ring made from a direct casting on a dried banana peel

Codina Goldsmiths. Design made from a direct casting on a geranium leaf.

3-D Prototype

Current computer technology offers tremendous possibilities for constructing designs, representing them graphically, and rendering them. There are various programs for three-dimensional creation, systems that make it possible to make elaborate patterns for jewelry on the computer. This technology saves time in constructing the design and making it, and it allows making designs that would be impossible by any other means.

All these systems complement one another with the new three-dimensional prototyping technology, especially lathe-cutter tools and three-dimensional wax printers. This technology makes it possible to transform a computer file into wax or resin for casting and to directly produce a mold.

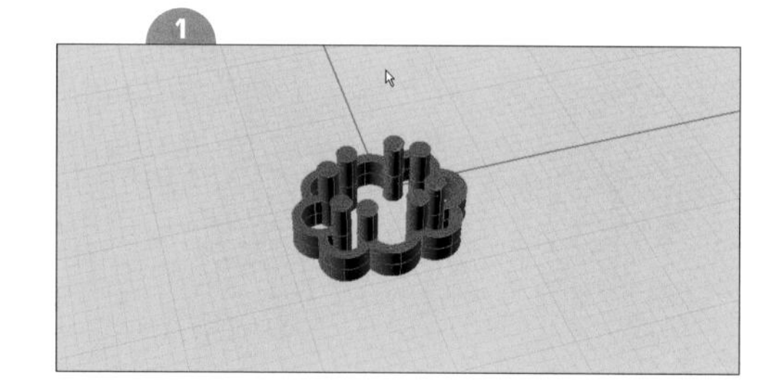

1/ File created with the Rhinoceros 4.0 program

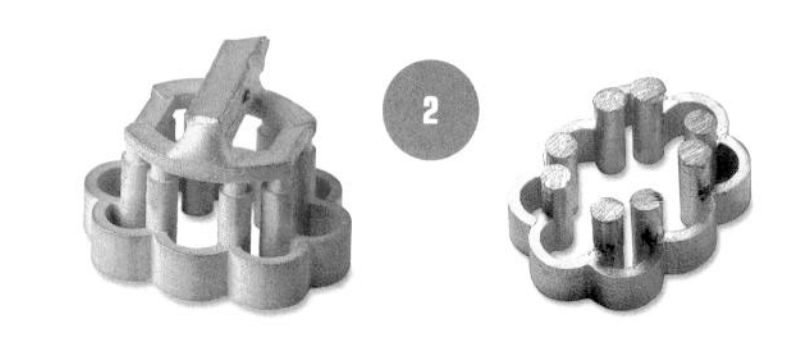

2/ The same piece printed in wax and cast in gold

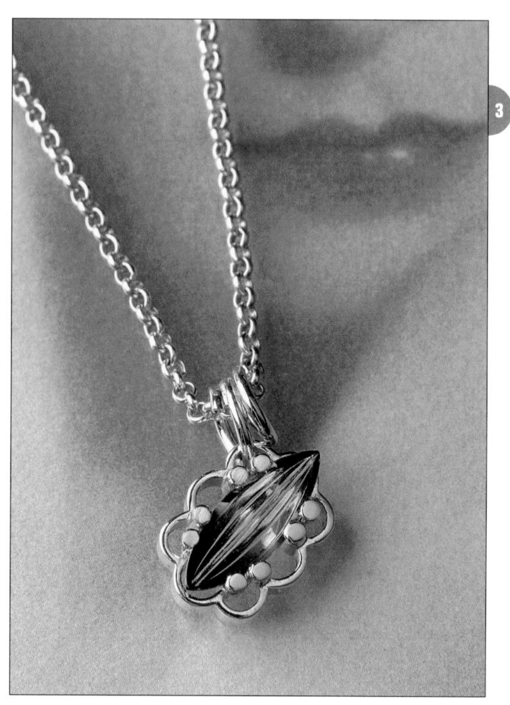

3/ The initial file has ultimately been converted into a pendant set with a tourmaline cut by Bernd Munsteiner.

Joan Codina. Pendant made from a 3-D pattern

Prototyping technology is being used more and more in jewelry, and not just for fabricating and casting designs. It can be used to make various components and findings, such as mounts, parts of rings, settings, and clasps. In addition, it is possible to quickly alter the same structure to fit different gems or different finger sizes, plus make multiple variants of a single design by working from the initial computer file.

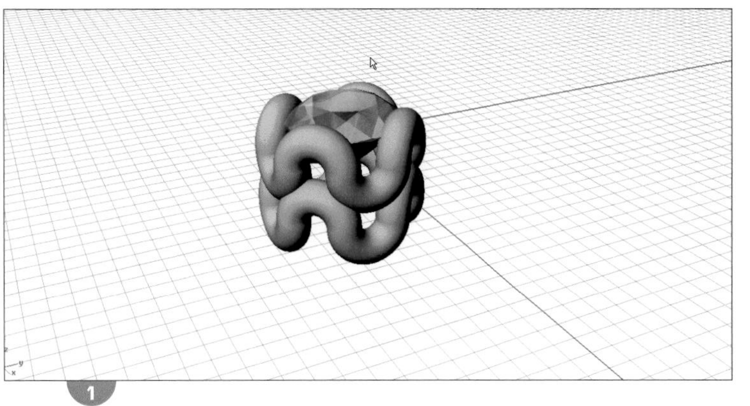

1/ To make a mount that fits various gem sizes, the scaling is done in the computer file, and then the pattern is printed in wax in the desired size.

2/ The pattern printed in wax will be microcast in gold.

3/ The casting sprues are cut off, and the texture is filed and finished with emery paper.

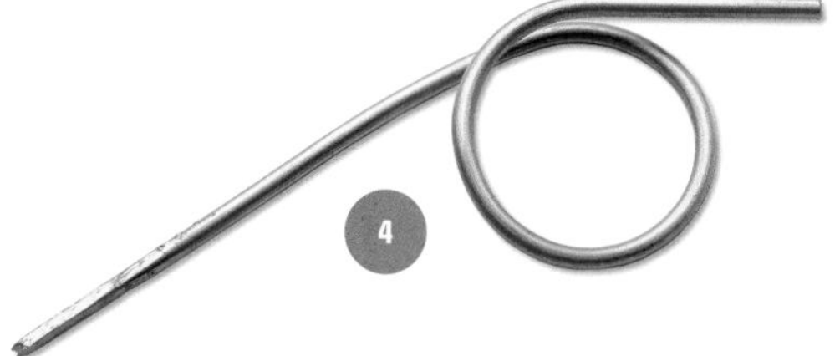

4/ A round wire that is 2 mm in diameter is wrapped around the ring mandrel, soldered, and rounded.

5/ After making a small fitting, the band is soldered to the gemstone setting.

6/ Since this process involves a computer file, the setting can be scaled to different dimensions and varied in numerous ways. This makes is possible to make different pieces and create collections.

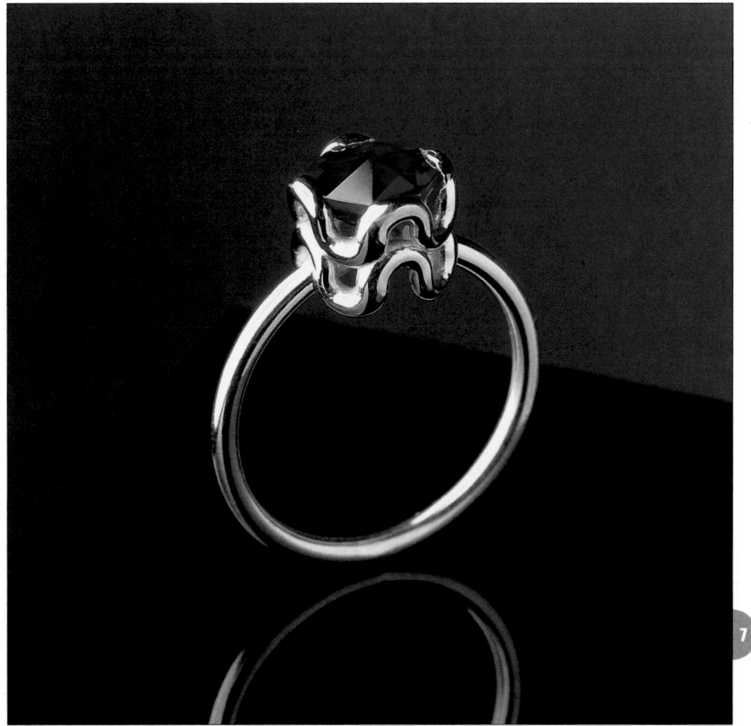

7/ The polished ring with a garnet set in the mount. **Codina Goldsmiths**

Modeling with a Lathe/Milling Machine

A precision lathe and milling machine is capable of producing a wax design by milling it from a compact bar or a thick sheet of rigid modeling wax. Once the file is loaded in the proper format, and after adjusting the various chucks and tools on the machine, it cuts and mills directly on the wax.

Quick prototyping machine from the Roland company with associated software

Close-up of a wax model being milled

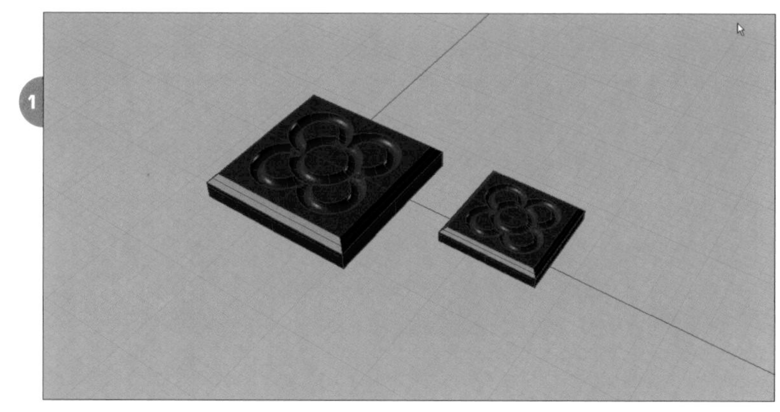

1/ Computer file created with the Rhinoceros 4.0 program

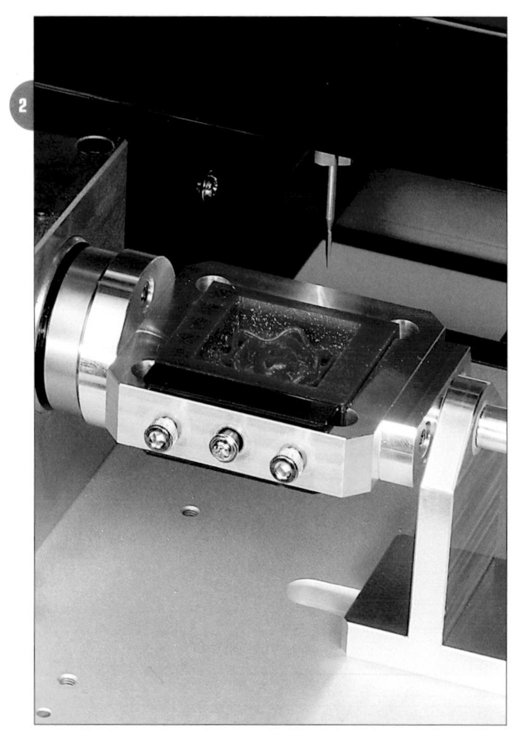

2/ This type of automated machine cuts directly on the rigid wax. Although it has limitations regarding size, it provides smooth, high-quality finishes in the milled areas.

3/ Various patterns milled in a block of green rigid wax

4/ After releasing the wax from the plate, small protuberances are removed from the design.

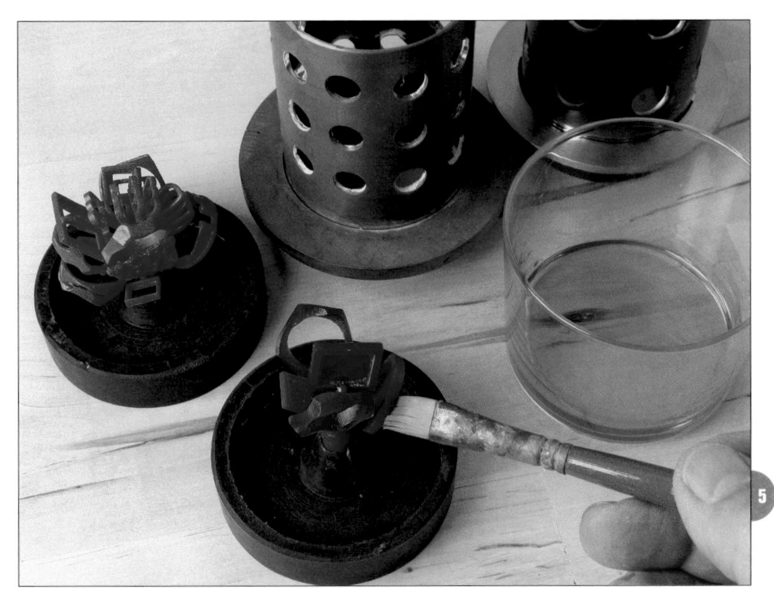

5/ The designs are attached to the wax tree and put into the coating cylinder. Then, they are burned out in the furnace.

6/ Once the design is cast, it is polished with the finest paper available. A metal wire is soldered onto the cast metal for making the vulcanizing silicone mold.

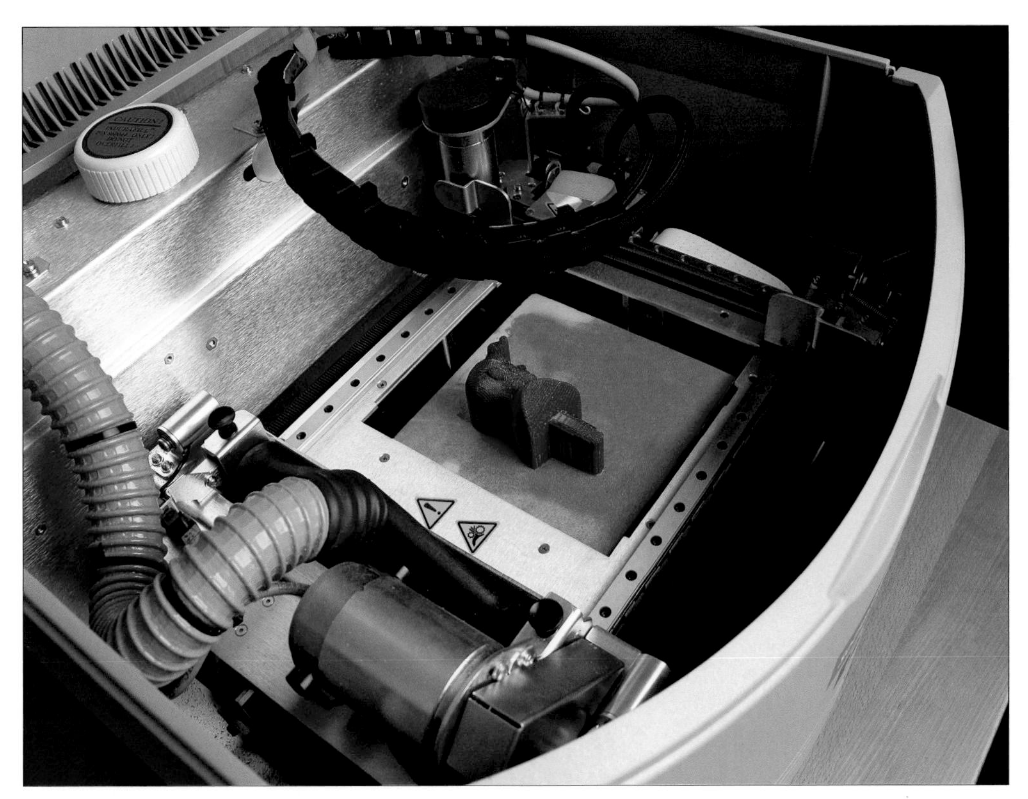

A completed print on a wax printer

Design with a Printer

A wax printer prints in three dimensions. It applies fine layers of wax through two heads: one of them prints a rigid blue wax, and the other prints a soluble red wax that serves as a form to support the blue wax. Later, this design will be microcast in metal.

Once the printing is done, all that remains is to remove the red wax so the blue wax can be cast.

Sand Casting

This is one of the most ancient casting techniques in existence. It is still used today for casting certain large-format designs in metal, especially sculptural pieces. Casting sand is easy to compact onto a design, keeps the shape, and allows the gases generated while pouring the molten metal to escape.

Using sand casting for jewelry is very practical. The technique is especially well suited for certain geometric designs. It produces the broadest array of objects in the shortest amount of time and yields a very interesting texture.

1/ The essential material is a type of finely sifted sand impregnated with the right proportion of burned oil to serve as an agglutinant. It is available from any jewelry supplier or from sculpture supply companies.

2/ You can buy various frames for sand casting, but it is very easy to make your own frames from wood or to modify a pair of picture frames so they fit together perfectly.

3/ Any hard item can be used as a design. For this example, we will make a design from machinable sheet stock, a material that is easy to file and rigid.

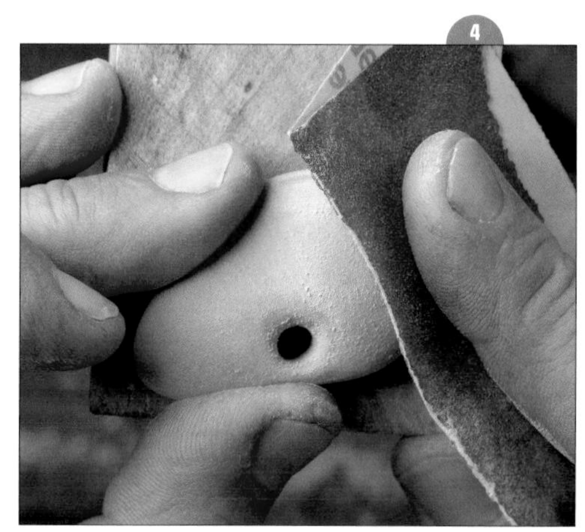

4/ Once the desired shape is completed, the surface of the machinable sheet is smoothed with emery paper down to 1200 grit.

5/ A square steel frame is selected and completely filled with compacted sand, first using the fingers and then a plastic mallet.

6/ The surface of the sand is smoothed and the excess is removed.

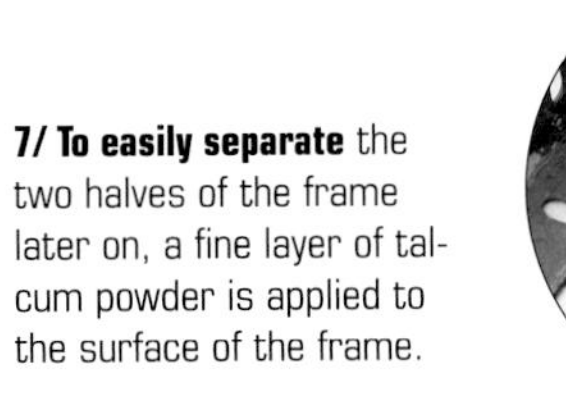

7/ To easily separate the two halves of the frame later on, a fine layer of talcum powder is applied to the surface of the frame.

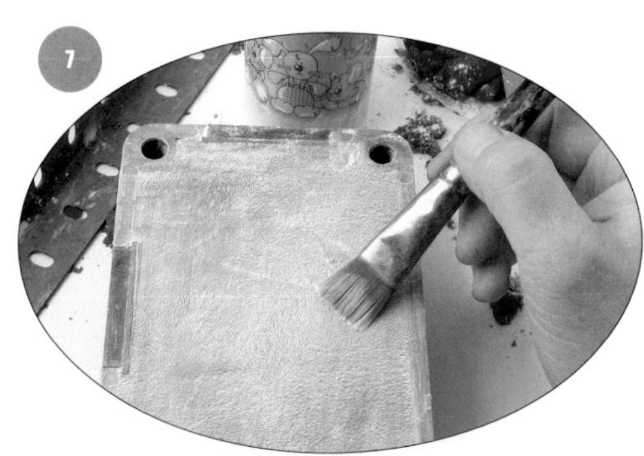

8/ To insert the design, it is sometimes necessary to remove a little of the sand, and then press the design in with your fingers. If necessary, you can lightly tap it halfway in.

9/ The second frame is attached using the indexes provided.

10/ The same process is used to fill the second frame with sand. It is compacted thoroughly with your fingers, and then with a mallet.

11/ The two frames are separated very carefully and the designs are removed. The runner gates are created to serve as feeders, and the pouring cup is created. Both must be wide enough to be filed with metal.

12/ The mold is closed once again, and enough molten metal is poured in the mold to completely fill the cavity, the feeder gates, and the pouring cup.

13/ Once the mold is opened, the cast silver designs will have to be cleaned and pickled. The runners are cut off, and the pieces are finished appropriately.

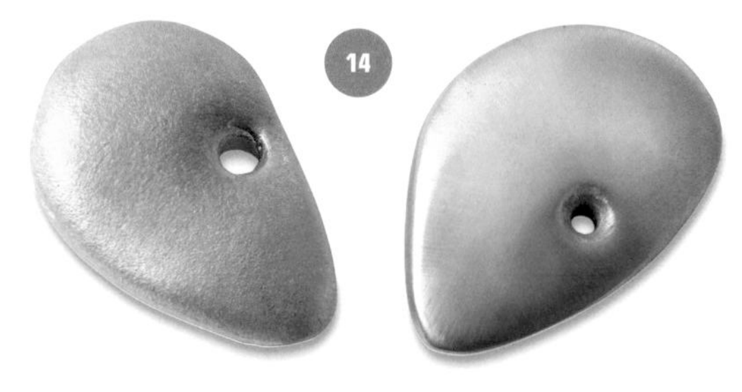

14/ The texture of the sand has been preserved on the piece on the left. The surface on the right has been filed and polished.

15/ A sand cast and polished metal keychain.

Cuttlebone Casting

Casting with cuttlebone is very practical in a small metal workshop. It is a quick and economical procedure that makes it possible to create a completely new object with unique characteristics. The essential material is the skeleton from a cuttlefish, normally discarded by the fishing industry. Molds from cuttlebone can withstand the high temperature of molten metal. The results are surprising. Like all materials, cuttlebone has its limitations, but it is easy to make copies of simple, rigid items, or to create objects by sculpting directly in this material.

The cuttlebone is the interior part of a mollusk called a cuttlefish. Similar to a skeleton, this product is considered waste and costs very little at the fish market. Cuttlebones should be completely cleaned and to left to dry before using.

Characteristics

With this casting method, it is possible to reproduce objects made of plastic, steel, wood, and any other material of sufficient hardness. The object must be able to be removed easily from inside the two halves of cuttlebone without damaging the mold. Casting with cuttlebone allows only a single copy of adequate quality. It is essential to make a new mold for each desired reproduction.

Sometimes, to create fine surfaces in metal, it is preferable not to clean the surfaces pressed into contact with the design and to coat this inner surface with a little liquid graphite.

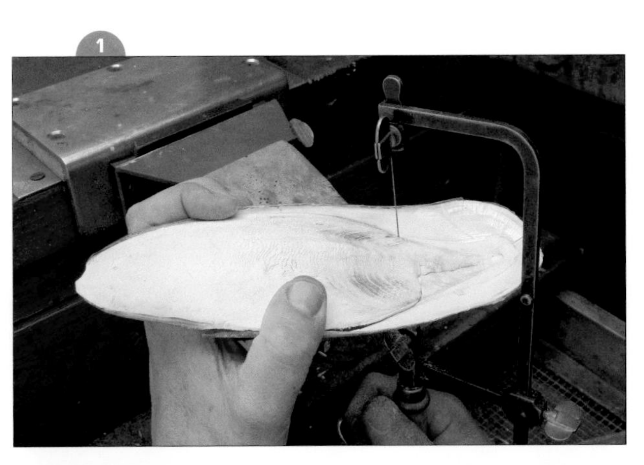

1/ After cleaning, the outside of the cuttlebone is cut off, leaving only the center, which is the spongiest area.

2/ A file is used to work down the central area and create a flat surface.

3/ A saw is used to cut through the middle of the cuttlebone, and then the two faces are sanded on fine emery paper.

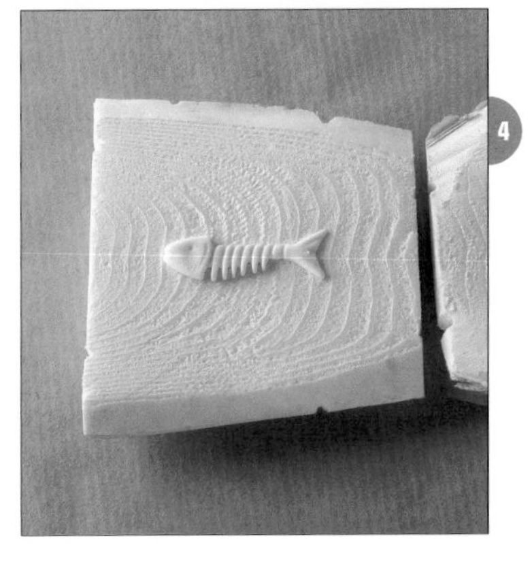

4/ A hard object is placed in the center of the two halves of cuttlebone. Two small pieces of bent sheet metal are used to make V-shaped indexing keys.

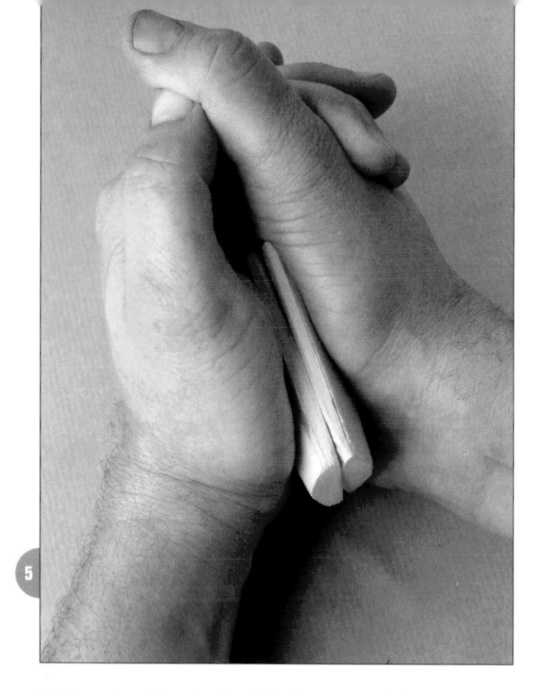

5/ Once the object is placed in the center of the spongy part of the cuttlebone, both halves are pressed together forcefully and evenly until they come in contact with each other once again.

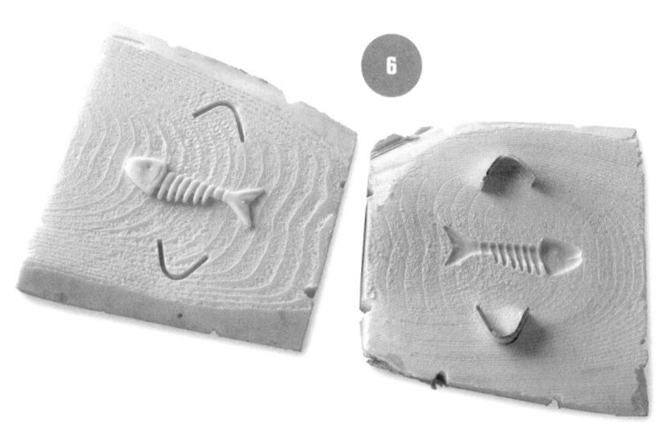

6/ The two halves are separated very carefully, and the object is removed. The metal indexing keys are left in place in one of the cuttlebone halves.

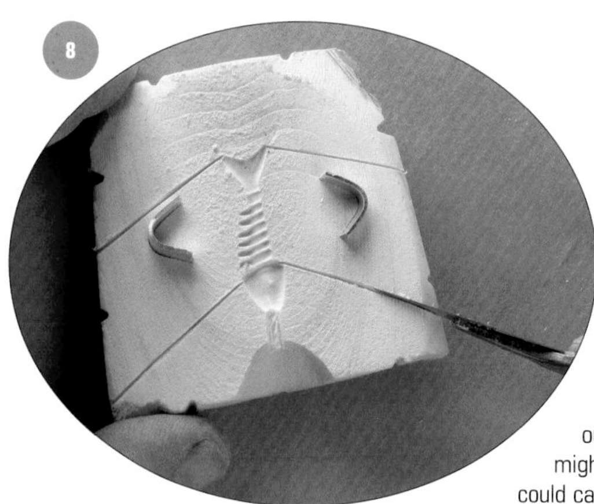

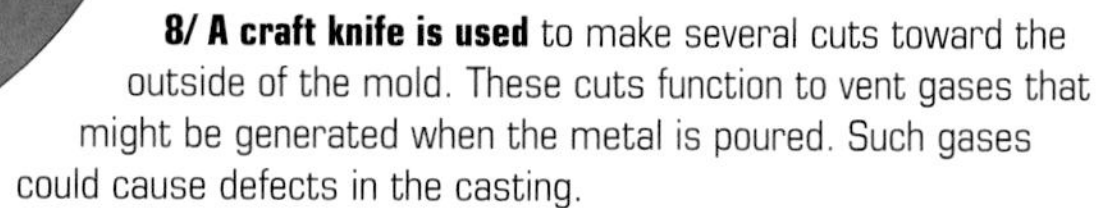

7/ A conical cut is made with a craft knife in the top part of the mold. It will be used for pouring the metal into the mold. This opening is always made near the thickest part of the design, and must be as close to it as possible.

8/ A craft knife is used to make several cuts toward the outside of the mold. These cuts function to vent gases that might be generated when the metal is poured. Such gases could cause defects in the casting.

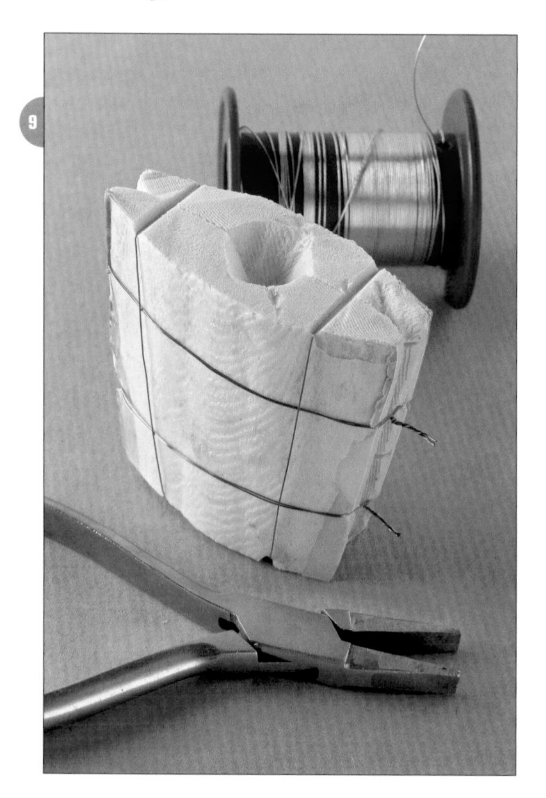

9/ The halves of the mold are fitted together using the indexing keys made earlier. They are then bound together with steel wire along the notches made previously.

10/ A quantity of sterling silver is prepared, more than the amount required. The cuttlebone mold is attached to several refractory bricks, and the metal is melted.

11/ The metal must be very fluid at the time of the pour. The cast is made quickly and precisely, from about 1 cm above the mold.

12/ The metal is allowed to cool. It is cleaned with a brush and water before being pickled.

13/ This is what a cuttlebone casting looks like once the pieces are cleaned. Note the size of the sprue and the texture produced. These proportions must be maintained to assure proper penetration of the metal inside the mold.

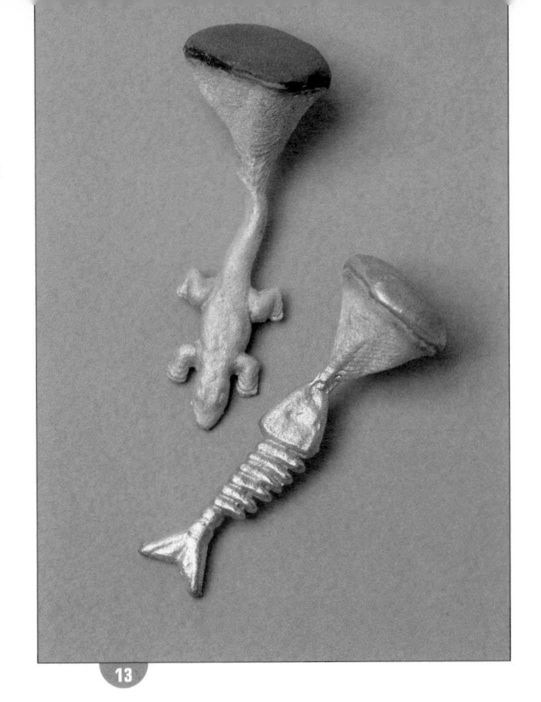

Original Designs

Original forms can also be cast without using an existing object to make an impression. Brushing or carving the bone itself yields interesting textures. Each casting is different, which gives each piece a special interest.

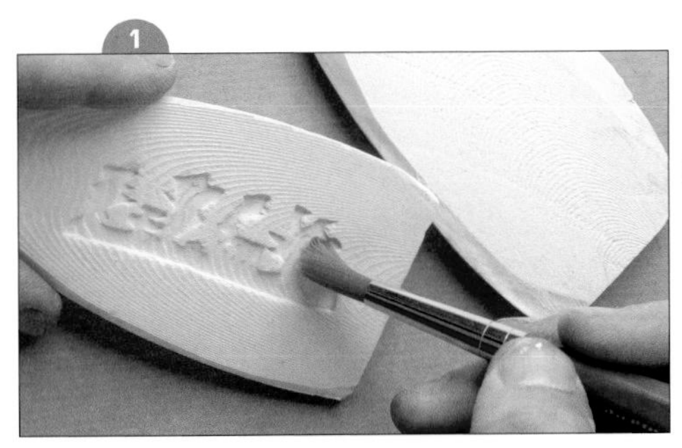

1/ The texture of the cuttlebone is interesting in itself. The surface merely needs to be brushed with a fine brush to make the texture of the material emerge. The mold is prepared in the way previously described.

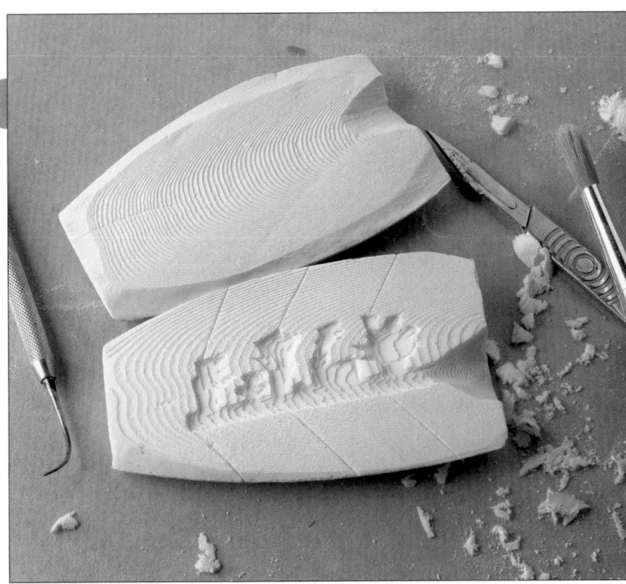

2/ It is possible to create forms by working directly on the cuttlebone. The surface can be cut or sculpted with a palette knife or objects and shapes can be pressed right into the soft interior of the cuttlebone.

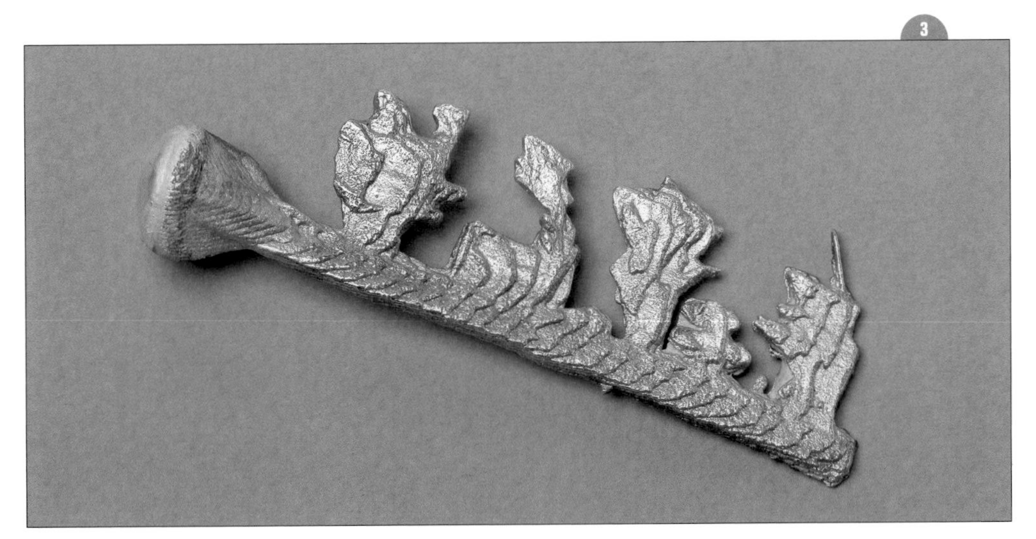

3/ This is the finished casting in silver. Note the texture produced by the cuttlebone.

Ceramic Coatings

Coatings can be made with very diverse materials. Among the most frequently used ones are many ceramic varieties. These are generally used in industrial casting processes and for casting sculptural objects, especially in alloys of bronze. The method that the author presents here is an adaptation of this casting method for jewelry. It makes it possible to cast small silver objects or small quantities of bronze. The benefits of this method include the quick preparation of the crucible and the good surface definition of the castings.

Preparing the Crucible

Making the wax base where the different pieces to be cast will be attached is easy. In this example, we have used a plastic holiday ornament. An amount of injection wax equal to about half the volume of the ornament is melted. This wax is poured into the plastic ornament, and it is closed. Next, the ornament is submerged in cold water and kept in constant motion. The cold causes the outside of the wax to cool down quickly. The result is a hollow wax shape. A 1-cm hole is made in the wax shape. The casting metal will enter through this hole once the coating has been applied.

1/ The injection wax with a low melting point is melted in a clean pan, taking care to avoid overheating it.

2/ The wax is poured inside a holiday ornament, preferably in the shape of a pear or a sphere.

3/ Several rods are made from the same injection wax. They will provide the proper feed to the designs once the wax is melted. The designs to be cast are prepared in the same way.

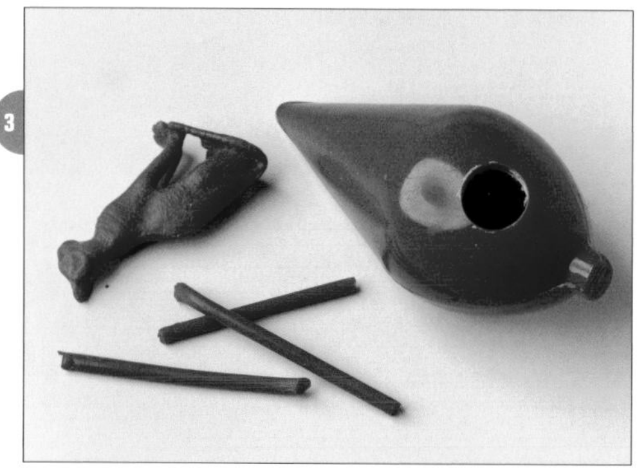

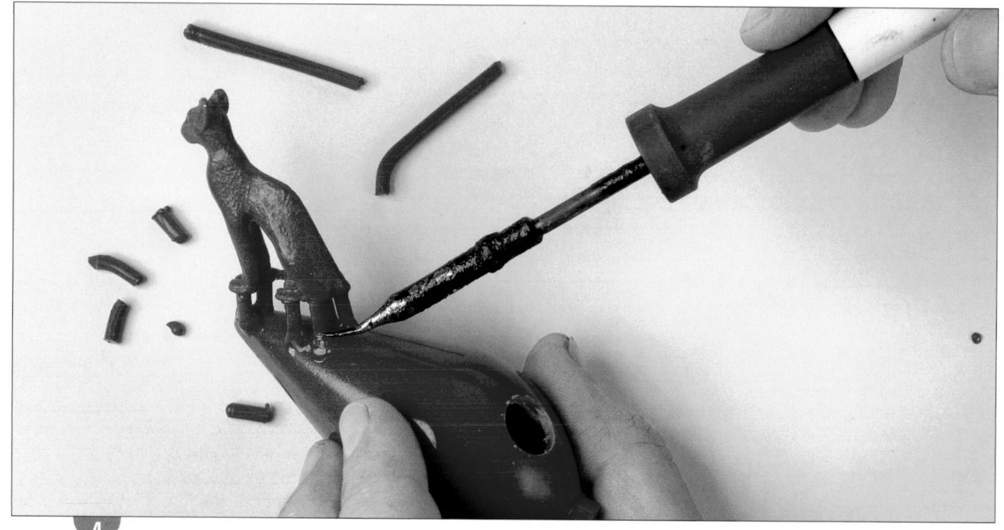

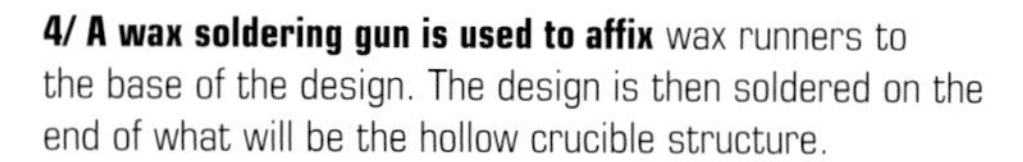

4/ A wax soldering gun is used to affix wax runners to the base of the design. The design is then soldered on the end of what will be the hollow crucible structure.

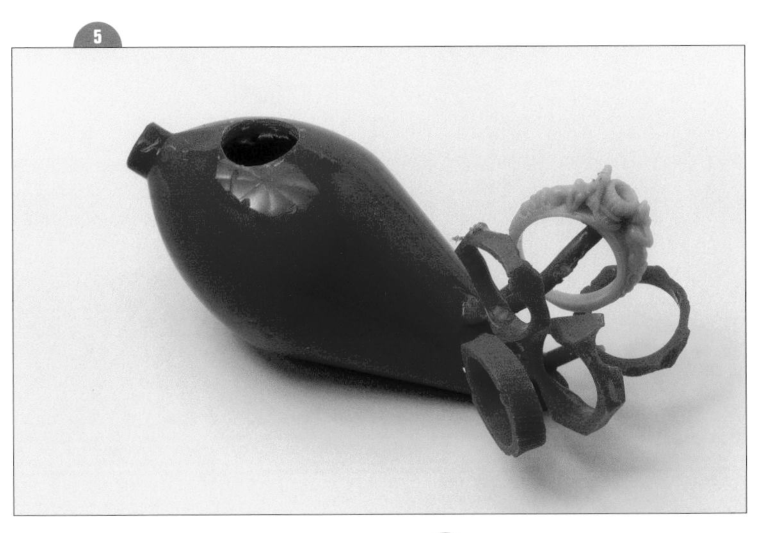

5/ It is possible to cast multiple pieces of wax at the same time. To do this, the various designs have to be soldered to the end of the crucible so the runners are as short as possible.

6/ It is possible to cast both wax designs and small plastic objects at the same time.

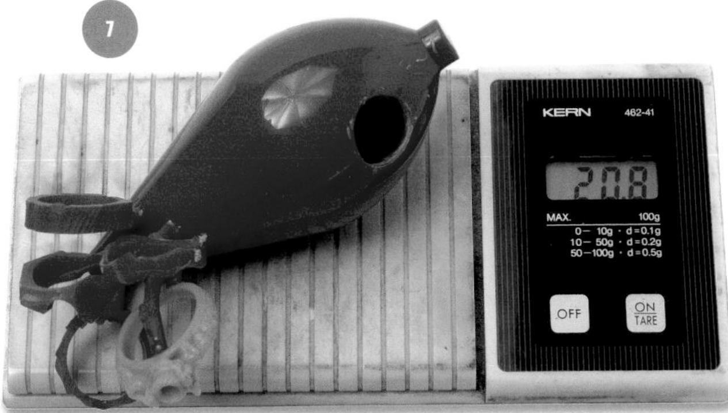

7/ It is crucial to know the final weight of the crucible with the wax or plastic designs attached. This helps you figure out the amount of silver or bronze that will be required. As a general rule, and always depending on the size of the wax crucible made, multiply the weight of the wax by 20 to arrive at the weight of the silver or bronze needed.

Preparing the Coating

Once the wax structure is prepared and weighed, the various types of coating are prepared. They will cover the wax and form a thick, heat-resistant layer that makes it possible to melt the metal inside the crucible structure.

First Coat

Before applying the first layer, the surface of the crucible is soaked with a thin coat of shellac diluted with alcohol. The purpose of this is to assure the perfect adhesion of the coating, which is composed of refractory clay, charcoal, and water. The wax design can also be painted with the shellac solution, but it is very important for this coat to be very thin. Too much shellac will cause the surface of the cast object to be rough and wrinkled. When the wax designs are very delicate, it is preferable to paint only the crucible and apply the first coat directly to the design.

1/ The shellac scales are dissolved in denatured alcohol to produce a liquid solution.

2/ The first layer of coating is prepared. It is composed of refractory clay and charcoal dissolved in water.

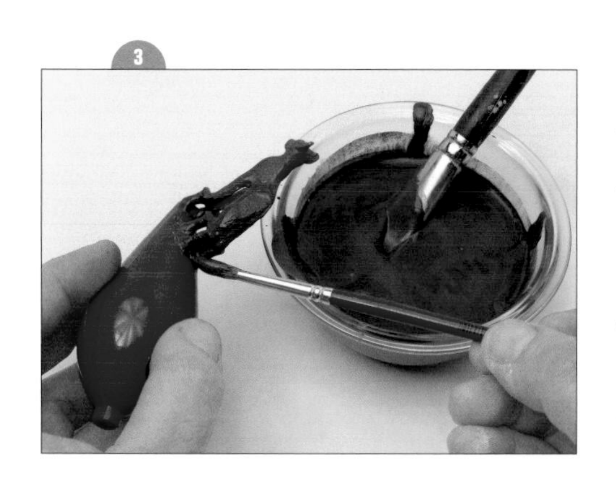

3/ A fine layer of the first coating is applied to all the wax designs with a flat, broad brush and allowed to dry.

4/ The surface of the crucible is painted with the shellac. The first layer of coating is applied on top of the diluted shellac.

Second Coat

The second coat is a mixture of sodium silicate, dark refractory clay, fine-grained malachite, and graphite. (Silicon carbide can also be used instead of graphite.) The sodium silicate acts as an agglutinant, and it also contributes temperature resistance. The refractory clay can withstand high temperatures. It must not exceed 30 percent of the total mass of the coating. Between 15 and 30 percent is an appropriate amount. The same volume of fine-grain malachite is added to provide strength. Finally, the charcoal is added, in half the amount of the malachite or clay. The second coat is applied and allowed to dry.

1/ A second coating is prepared from sodium silicate, refractory clay, fine-grained malachite, and very fine ground graphite. Everything is mixed and set aside.

2/ The second coating is applied with a flat brush to produce a layer that is around 2 mm thick. This thickness allows the adhesion of the medium-grain malachite.

3/ While this layer is still wet, the entire surface is sprinkled with medium-grain malachite and allowed to dry and harden.

4/ It is important that all areas be filled with malachite. Any imperfection in these first layers may cause cracks or holes in the coating and a resulting loss of metal in casting.

Third and Fourth Coats

The third coating is prepared in the same way as the second, but without adding graphite to the mixture. Once the paste is applied, the surface is dredged in a mixture of medium and coarse malachite. Everything is allowed to dry. A final layer of the same coating is applied, but in contrast to the previous one, this one is dredged only in coarse-grained malachite. The result is a good thickness and greater consistency. Once everything is dry, wax removal can begin.

1/ A coating composed of dark refractory clay and sodium silicate is mixed to the consistency of yogurt.

2/ The structure is very carefully submerged in the coating or it is painted on with a brush, producing a thick layer on the entire surface.

3/ A mixture of equal parts of medium- and coarse-grain malachite is prepared. The structure is dredged in it uniformly. The piece is set aside to dry completely.

4/ The preceding process is repeated for the fourth layer, but now only coarse-grain malachite is sprinkled on the structure.

5/ Once all the crucibles have dried, it is necessary to empty the wax so the metal can be poured into the resulting space.

Casting

Before casting the metal, we have to completely remove the wax from inside the crucible. It is then loaded with the precise weight of throughly clean metal.

Load the furnace with charcoal to unify the heat and reduce the oxidation as much as possible. Once the furnace is ready, the crucible is put into place and slightly tipped toward the thick side where the metal is. Using an enveloping flame that covers the entire crucible, the melting is begun. The advantage of this system is that the temperature spreads out evenly, contributing to proper melting.

1/ To remove the wax, hold the crucible vertically with tongs. Orient the hole for pouring the metal at the bottom. Apply heat directly to the crucible with a torch to eliminate the wax completely.

2/ Weigh the metal. Generally, if you want to cast silver, all you have to do is multiply the weight of the wax by 20 and put that amount of metal into the crucible. For gold or other alloys, the weight of the metal is increased based on the density of the metal to be cast.

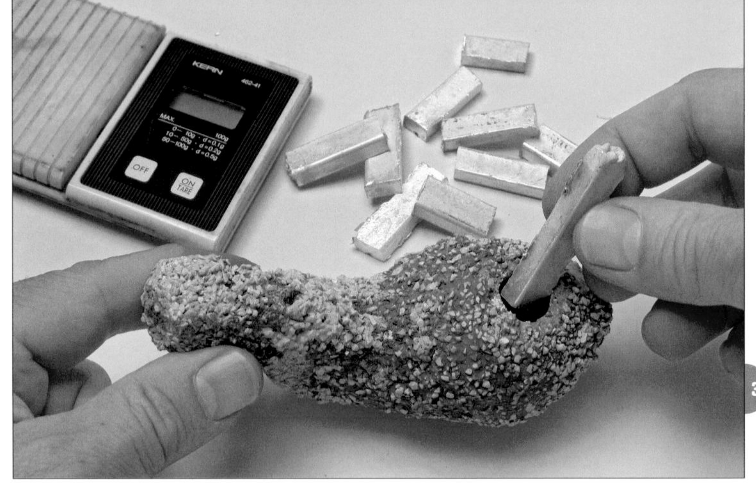

3/ For this example, we have chosen to melt silver. It has been cast into ingots and cut into sizes that easily fit in the crucible.

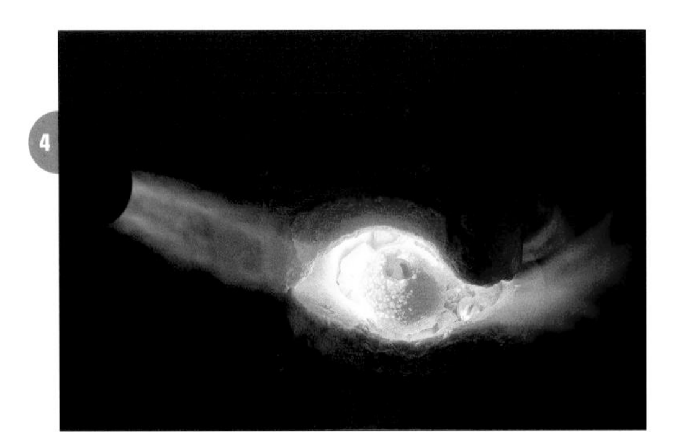

4/ A furnace loaded with charcoal is prepared, and the metal is melted inside the crucible. The crucible must be positioned in such a way that once the metal liquefies, it remains in the rearmost part of the crucible.

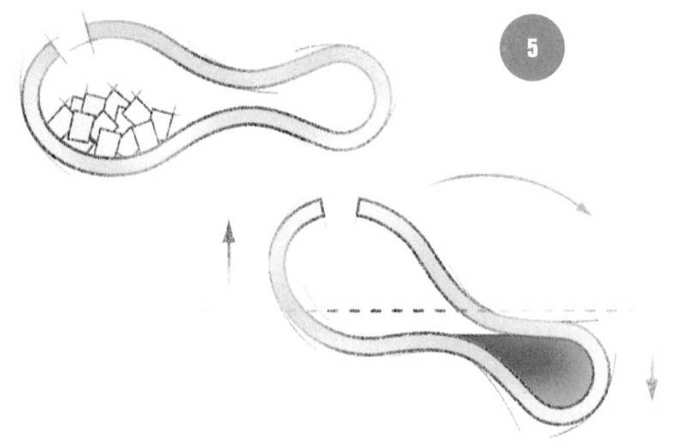

5/ The top drawing shows the position of the crucible inside the furnace. The bottom drawing shows the crucible tipped over as described in steps 7 and 8.

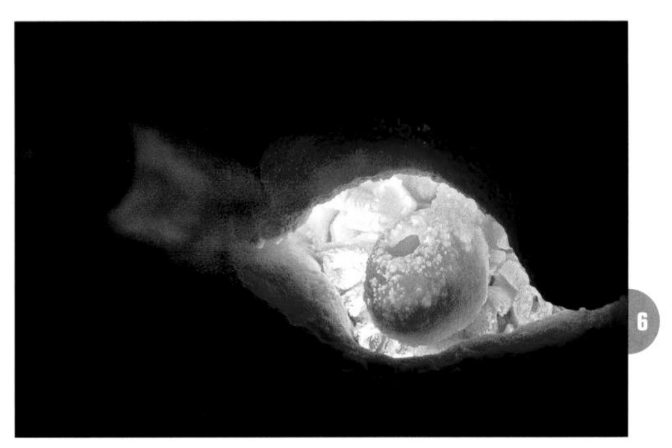

6/ The metal is melted with an enveloping and reducing flame that covers the whole piece to produce a liquid, fluid metal.

7/ The crucible is grasped with tongs and tipped so the liquid metal quickly flows into the end and fills the space left empty by the wax.

8/ The crucible is left in this position until the metal solidifies. When it changes its initial red color to a brownish-black, the crucible is taken out of the furnace and cooled in water.

9/ The coating is struck with a hammer, taking care to avoid damaging the metal pieces.

10/ It is best to remove as much coating as possible before sandblasting the surface of the cast metal.

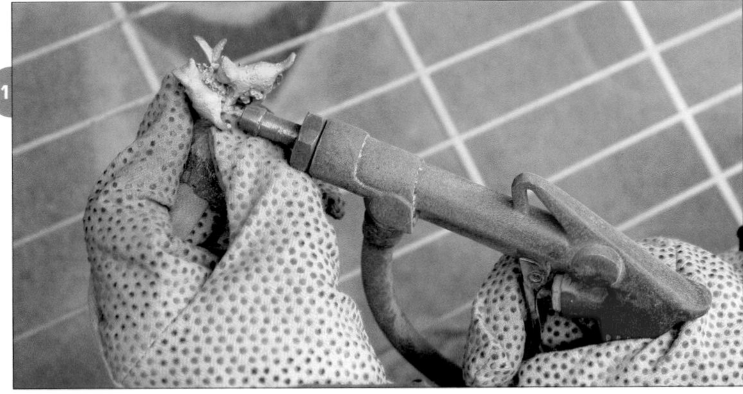

11/ A sandblaster works well for removing coating residue and producing a smooth, refined finish.

12/ Finally, the castings are put into a 20-percent solution of sulfuric acid and water to whiten the metal.

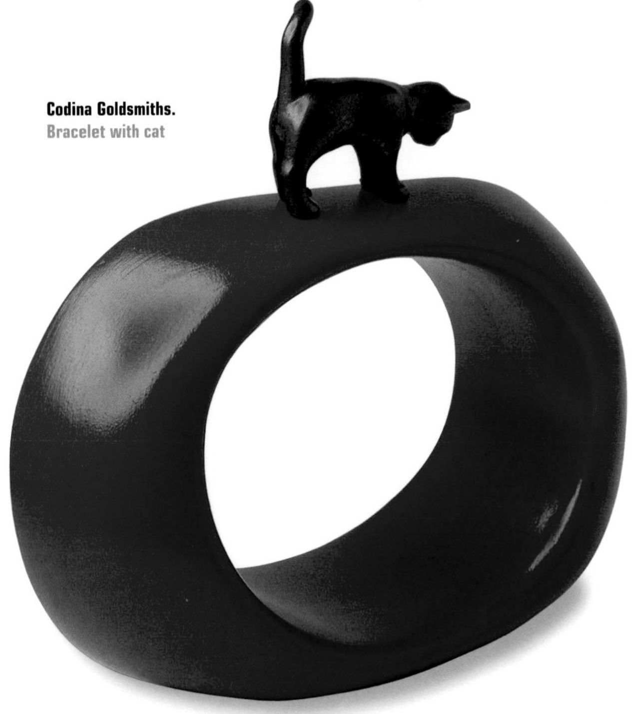

Codina Goldsmiths.
Bracelet with cat

Casting in a Closed Crucible

This technique is a variation on an ancient method used in China and still used in Africa for casting small sculptural pieces and jewelry. It is a system very similar to that currently used by the Ashanti tribe in Africa. This version introduces some new materials used in the preparation of the coating and in the construction of the furnace.

Closed crucible casting basically involves constructing a crucible with the metal already inside it, along with a wax design. The wax is removed and the casting is made in a furnace. By simply tipping the crucible, it is very easy to introduce the molten metal inside the hollow space left by the wax.

The main advantage of this closed procedure is that the temperature of the crucible and the metal are perfectly equal at the time the metal is cast. This allows a perfect, easy pour. At the same time, this coating, in combination with its special closed design, ensures a significant presence of carbon during the process, which guarantees a proper cast.

Preparing the Crucible

A foam sphere is cut in half. The wax designs to be cast are attached to the surface of one of the foam halves. Feeders, in the shape of wax rods, are soldered to the designs in advance to ensure proper casting. It is preferable to use one wax design per crucible. This makes it easier to apply the coating, which must be done with your fingers.

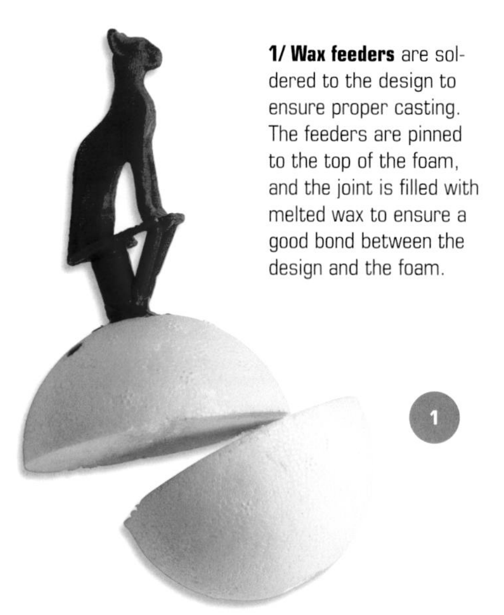

1/ Wax feeders are soldered to the design to ensure proper casting. The feeders are pinned to the top of the foam, and the joint is filled with melted wax to ensure a good bond between the design and the foam.

The design can be made from any type of modeling wax, and it is also possible to cast some plastics directly. It is very important to weigh the wax correctly before connecting it to the foam sphere, and to multiply this weight by 22 to determine the weight of the metal.

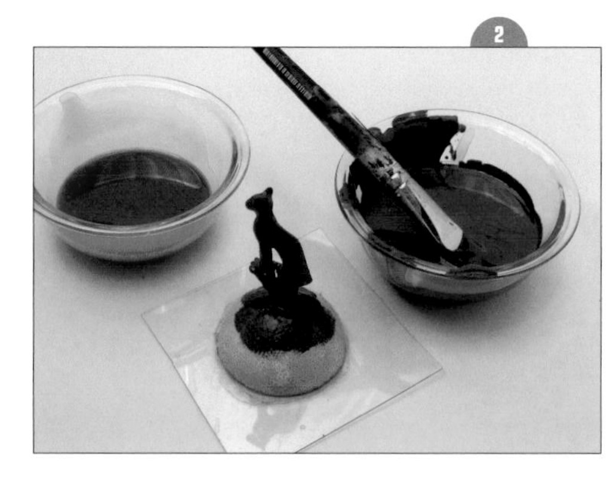

2/ The entire surface is painted with shellac dissolved in denatured alcohol. This layer must be very thin. Otherwise, it will produce a wrinkled, rough texture once it is burned out in the furnace. Then, the first layer of coating is applied. It is made of potter's clay and charcoal in equal proportions.

Coating

Prepare a very thin mixture of 50-percent very fine powdered charcoal and 50-percent potter's clay highly diluted in water. This coating must remain liquid and homogenous when applied with a

1/ To prepare the thick coating, high-fiber, whole-grain cereals are mixed with natural esparto grass (cut very fine with scissors) and a little water.

2/ The mixture can be ground or crushed in a mortar, adding a little water as needed to homogenize it.

brush over the layer of shellac. Let this dry for at least 10 hours.

Then, start to apply the coating in two varieties—a first, finer coating with charcoal content, and a second, thicker one without charcoal. Both coatings must remain homogenous and well compacted. To apply them, press the coatings very carefully around the design, creating a final thickness of at least 1 cm.

3/ Prepare the first coating with 40-percent charcoal, 25-percent potter's clay, and 10-percent grog or refractory clay. Mix the ingredients slowly by hand to produce a compact paste.

4/ The paste must be homogenous and have the right consistency and fineness to adhere properly to the surface of the design.

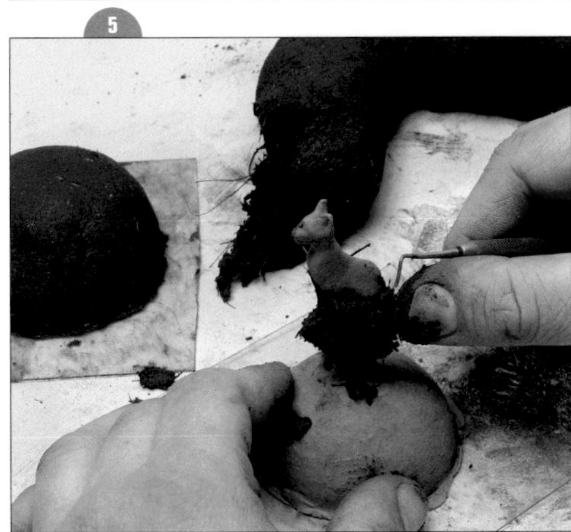

5/ Press with your fingers and a palette knife to cover the design, making sure to fill all the little inner spaces.

6/ The same coating is used to cover the crucible to a thickness no greater than 7 to 8 mm.

7/ For the second coating, the cereal-and-esparto-grass mixture is combined with an equal amount of potter's clay and 30-percent grog.

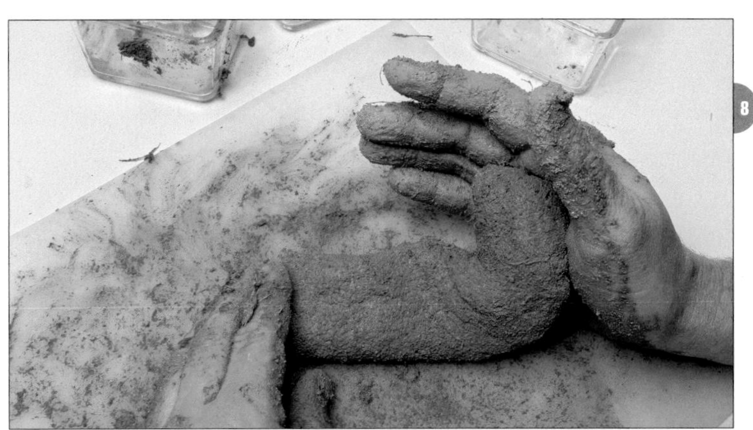

8/ These ingredients are mixed to produce a homogenous paste.

9/ A layer that is slightly more than 1 cm thick is applied to both sides of the crucible. It must be compressed with the fingers so that it is perfectly compacted.

10/ Let the coated crucible parts dry for several days. Fill any crack that appears with a slightly more liquid form of the final coating, and let the pieces dry once again.

11/ The foam is removed. The two halves of the crucible are smoothed so they fit together.

12/ Tongs are used to hold the piece of the crucible that contains the pattern. A flame is applied directly with the torch to burn out the wax inside.

13/ The weighed metal is put inside the crucible. (Sterling silver was used in this sample.)

14/ The two pieces are joined with a small amount of the first coating. It is moistened and applied to the two halves, and then they are squeezed together.

15/ A finger is moistened with water, and the whole piece is refined by hand. It is allowed to dry completely before being put into the furnace.

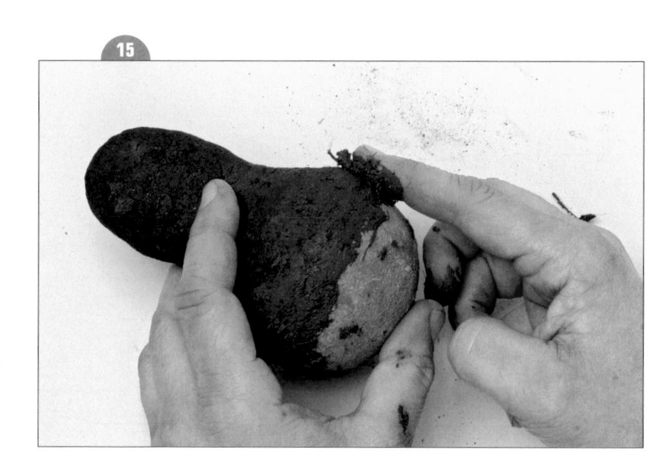

Constructing the Furnace

Next, we will show how to construct a very simple furnace for casting with closed crucibles or ceramic molds. This is a vertical furnace fueled by oak charcoal with the addition of coal to increase the temperature. Coal produces more heat than charcoal, so the temperature will be regulated by the proper quantities and the quality of the coal.

The furnace will take three or four hours to reach the casting temperature, so it's a good idea to prepare many crucibles with different designs, and put them all into the furnace in succession to make a production run.

The furnace should always be constructed keeping in mind the number of crucibles to be cast. The photo shows a large-capacity furnace fed by two large pipes connected to their own air supply.

1/ To make a small furnace, a base is first made with fine refractory bricks on which the walls of the furnace will be raised.

2/ Thick white refractory brick is used to construct the walls of the furnace. It is important to leave some 5-cm openings between the bricks on the bottom. This lets sufficient air into the furnace and produces the right air convection.

3/ The walls are stacked to a height of three or four bricks. If there are quite a few crucibles to be cast, the furnace can be enlarged up to six bricks high.

4/ Cut a strip of thermal blanket and wrap it around the furnace to prevent heat loss.

5/ Tie the thermal blanket in place with steel wire, and fit it to the walls of the furnace.

6/ Oak charcoal and a little coal are put into the furnace.

7/ The furnace is covered with a piece of thermal blanket. Make sure that the openings in the lower part of the furnace allow the air to enter.

8/ The furnace should turn orange in color, indicating that it has reached the right temperature.

9/ The crucible is put inside the furnace in a vertical position. It is covered with charcoal until it returns to temperature. The metal inside the crucible is melted.

10/ To check if the metal is melted, use tongs to pick up the crucible and tilt it to the side. If you detect movement of the metal inside the crucible, then it is in a liquid state.

11/ The crucible is quickly turned 180 degrees on top of the furnace. This moves the molten metal into the space left by the wax pattern.

12/ The crucible is left in this position until everything cools down. It is opened by giving the joint a sharp rap.

13/ When the metal is completely cool, start breaking off the coating, which has hardened because of the heat.

14/ The rest of the coating is removed, taking great pains to avoid striking the cast metal.

15/ Though you can remove all the coating by cooling the piece abruptly in water, it is better to use a sandblaster to clean it completely.

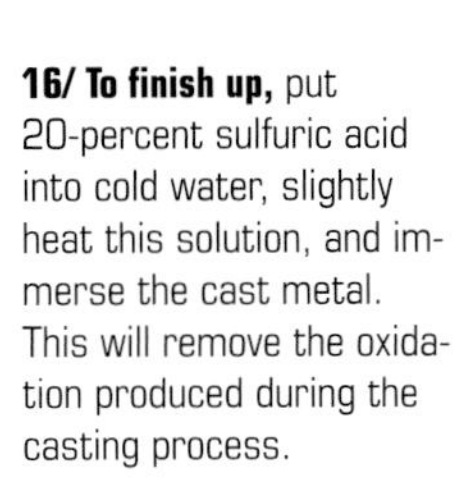

16/ To finish up, put 20-percent sulfuric acid into cold water, slightly heat this solution, and immerse the cast metal. This will remove the oxidation produced during the casting process.

Codina Goldsmiths. Finished pieces made with this process

Cast Bronze Bracelet

The following project shows how to make cast bracelets in bronze and to patina them using different methods. Few tools are needed to create the patterns in wax.

For this project, red wax, which is normally used for mold injection, and a piece of ceramic clay are needed.

1/ A piece of stoneware clay has been properly kneaded and smoothed out. The wax is impregnated with a fine coat of silicone applied as an aerosol spray. Various shapes are pressed into the surface to create the desired patterns.

2/ The injection wax is melted in a pot at the lowest possible temperature, making sure that it does not boil. Then, it is poured onto the stoneware clay and spread out evenly.

3/ The wax is allowed to cool, then the clay is removed from the surface. A little vinegar can be used to completely remove the clay.

4/ Using the wax soldering iron and some craft knives and palette knives, the outside of what will become the bracelet is shaped. The inside is hollowed out to reduce the weight of the bracelet.

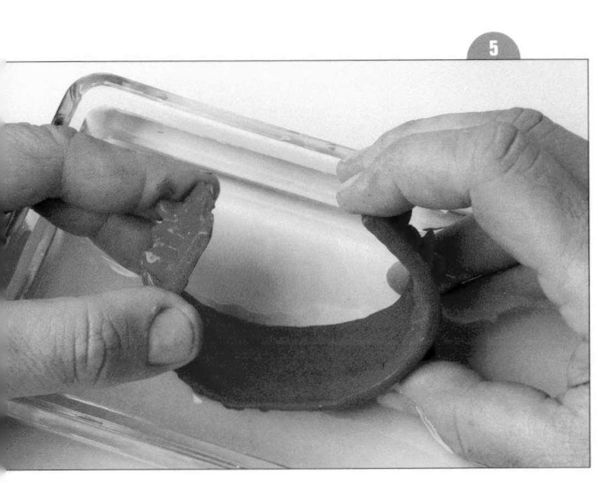

5/ To shape the bracelet, the wax is submerged in hot water that is about 86°F (30°C). The wax is very carefully bent to produce the desired shape.

6/ The wax is put into cold water to set the shape. A wax soldering iron is used to finish shaping and correct small errors.

7/ The wax bracelet is attached to a casting tree so the casting cylinder can be prepared. (The process can also be done by another method, or the wax can be brought to a casting specialist.)

8/ The bracelets are cast in bronze, but they could also be cast in silver or some other metal.

9/ Once cast and cleaned, the bracelets are patinated with potassium sulfide, iron sulfate, and titanium oxide to produce three different colorations.

Codina Goldsmiths.
Bronze bracelet

Codina Goldsmiths.
Bracelet made with the same process, cast in silver, oxidized with potassium sulfide, and polished

Sand Cast Silver Rings (Tensi Solsona)

Tensi Solsona shows how to make original designs by working directly with the casting sand. She creates a variety of different rings for which there is no pattern. To do this, Solsona assembles a variety of materials, prepares an assortment of tools for working with the compacted sand, and casts the metal with great precision.

1/ To make the mold, we need a set of aluminum frames. Here, we use two cylinders with indexing notches on the outside so the frame can always be put back together in the same position.

2/ First, the shallower aluminum cylinder is filled with sand and compacted with the fingers. A mallet is used to pound the surface and further compact the sand.

3/ A knife is used to level the surface of the sand.

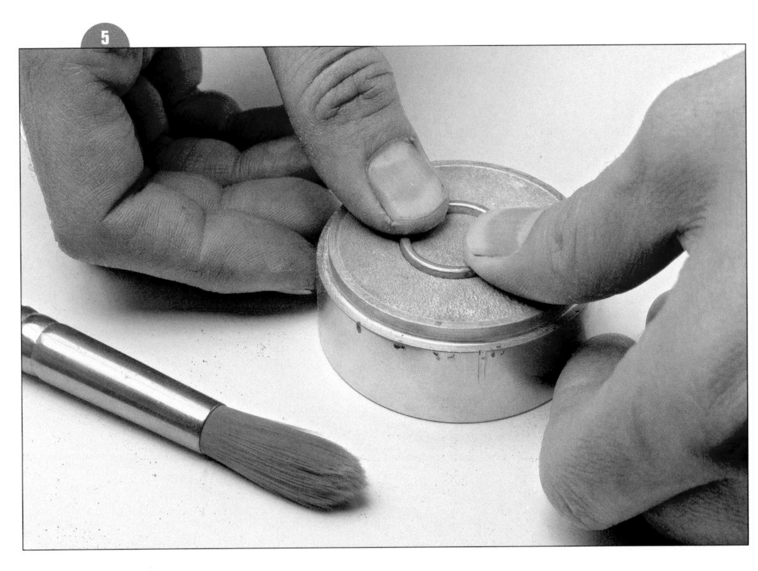

4/ Using a fine round brush, a thin layer of talcum powder is applied to the surface of the sand. The talcum powder will separate the two halves of the mold.

5/ To make the main body of what will become the ring, a metal circle is pressed halfway into the compacted sand.

6/ The second aluminum cylinder is put into place, making sure that the indexing notches on the perimeter match up perfectly. The second cylinder is filled and the sand is packed tightly inside it.

7/ The aluminum frame is separated with great care, and the metal circle is taken out. The sand is modified using a variety of steel rods and spheres. They are pressed in very carefully, and then removed to create the desired design.

8/ Once the imprints are made, the metal circle is put in the sand again. It is pressed slightly to correct any crumbling of the sand in this area. Perpendicular to the main body of the ring, a 3-mm rod is used to make a feeder for the metal.

9/ The metal circle is removed, and the second aluminum cylinder is put into place, making sure that the indexing notches line up. A knife is used to open the pouring cup on the surface of the second ring.

10/ Using a broad palette knife, the sand is pressed around this area to compress the sand to keep it from crumbling during the casting process.

11/ The sterling silver is melted. The framed sand is placed close to the furnace.

12/ The pour must be done quickly and precisely, from about 2 cm above the surface of the frame.

13/ Once the cast is made, the burned sand is removed. The clean sand can be reused later.

14/ The piece as cast, before it is put into a pickle solution.

15/ It is important to file the silver to remove any casting burrs and small defects that may have occured when the metal was poured. In particular, the flat parts of the metal that will be visible on the finished piece are filed.

16/ Rings cast in silver and oxidized with potassium sulfide.

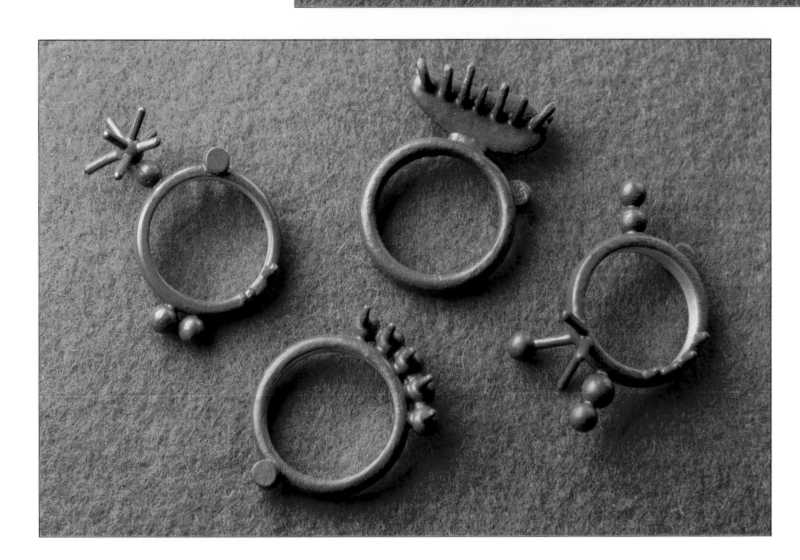

Tensi Solsona. Oxidized silver rings

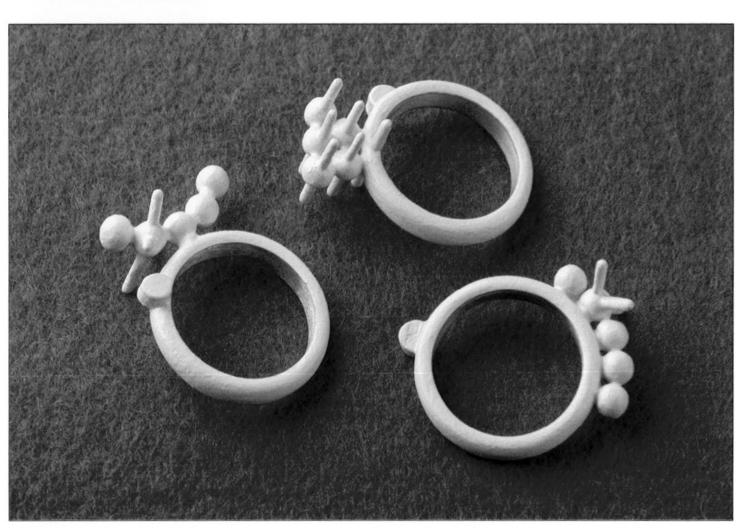

Tensi Solsona. Sand-cast rings

Cast Leaf Bracelet

In the following pages, I will show you how to make an articulated gold bracelet based on a geranium leaf. We will show the entire progression from the time the leaf is cut from the living plant to its transformation into a metal pattern and the subsequent casting of multiple copies. We will also show the entire finishing and fabrication process for a cast gold articulated bracelet.

It is important to point out the special characteristics of a cast piece. The casting process yields a more fatigable, less dense metal than laminated or drawn gold. As a result, it is necessary to avoid using cast metal for all elements that are subjected to wear, such as joints, clasp tongues, and even clasp hooks.

Although it is possible to do the entire process with a small casting setup, many jewelers regularly have all or part of it done by a casting professional. In any case, it is essential to be familiar with the process that a design will have to undergo to be transformed into a piece of jewelry.

1/ A leaf with an attractive shape is selected for the design. The leaf should not be very thin or excessively large, so the element is not too heavy when cast in gold.

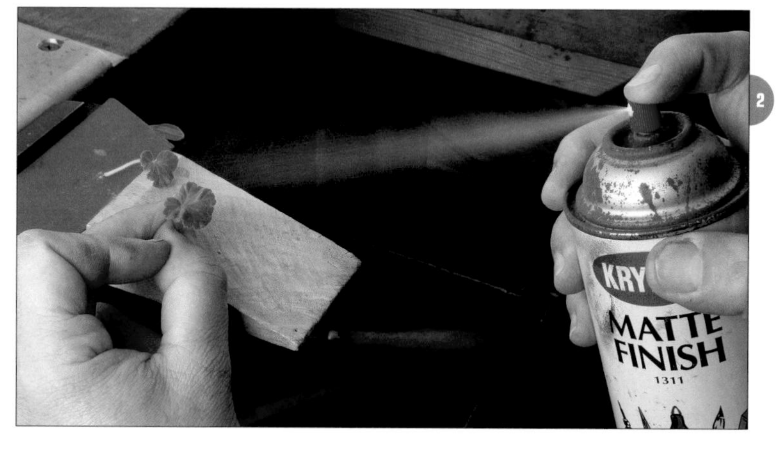

2/ To stiffen the natural leaf, the surface is sealed with a spray finish that does not contain tars. This avoids the potential for residue in the cylinder once the leaf is burned out.

3/ Using dental wax and a round brush, the entire inside of the leaf is reinforced to thicken it slightly. This keeps the leaf from losing its shape during the 10 minutes it will have to stay inside the liquid coating. This also causes the leaf to reach the thickness necessary for making a vulcanized mold.

4/ The next step is to cast the leaf directly in a cylinder, as if we were dealing with a wax design, and thus make a metal pattern. To ensure good burnout, the maximum temperature is increased to 1560°F (850°C) and kept there for at least one hour. Finally, a pouring cone is soldered to the cast and cleaned pattern, and it is ready for reproduction in a vulcanized silicone mold.

5/ A silicone mold of the metal pattern is made. The mold is cut through the middle and opened to retrieve the metal pattern and inject the wax.

6/ As many wax pieces as necessary are injected, up to the total required for the gold bracelet. As a standard measurement, we figure on a bracelet length of 18 cm.

7/ A thick trunk of injection wax is made. All the wax copies needed are soldered onto the wax tree.

8/ The wax tree is weighed, and the resulting weight is multiplied by 15.5, which is the density of the gold. An additional ½ ounce (15 g) is added to account for the pouring cone.

9/ The tree is attached to its rubber base, and a perforated steel cylinder is chosen for vacuum casting.

10/ The coating is prepared and poured into the cylinder. Then, it is put into the vacuum pump to remove the air inside the moist coating.

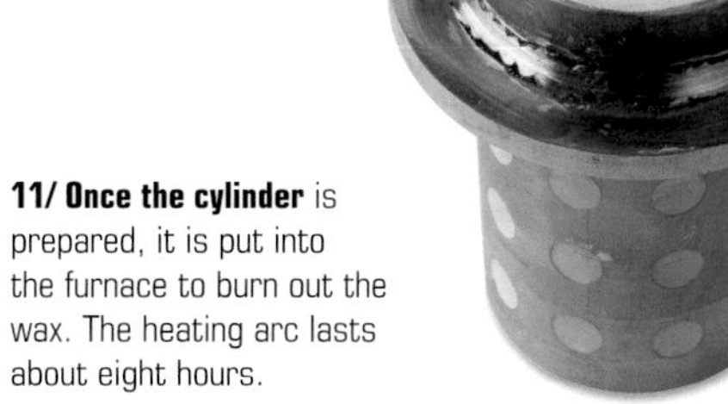

11/ Once the cylinder is prepared, it is put into the furnace to burn out the wax. The heating arc lasts about eight hours.

12/ The molten metal is poured into the cylinder, and the residue from the coating is removed. The 18-karat gold tree is pickled, and the assembly of the bracelet begins.

13/ A wire cutter is used to cut off the sprues. Then, a jeweler's saw is used to trim the sprue close to the outline of each cast piece.

14/ The gold leaves are cleaned carefully with files. Care is taken to avoid damaging the textured surfaces of the leaves.

15/ The leaves are matched up in pairs. To ensure a good solder bond, a triangular file is used to fit them together. Then, the pairs of leaves are soldered.

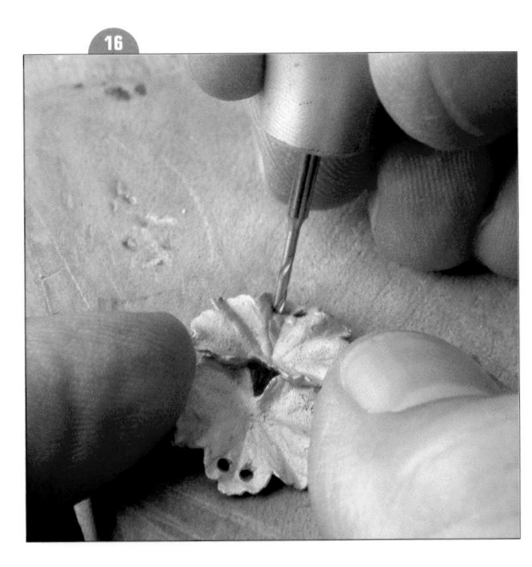

16/ Using a 1-mm bit, two adjacent holes are drilled in the ends of each pair of leaves.

17/ To produce a reliable and durable joint, we used drawn gold wire that was 0.08 mm in diameter. Using a cylindrical bit measuring 0.08 mm in diameter, a small space is ground between the two holes.

18/ The gold solder is applied to the ground space, and a piece of drawn wire is soldered on. This is so the strain of the joint falls onto the drawn metal and not onto the cast piece.

19/ As shown in the photo, several U-shaped pieces are made with the 0.08-mm wire.

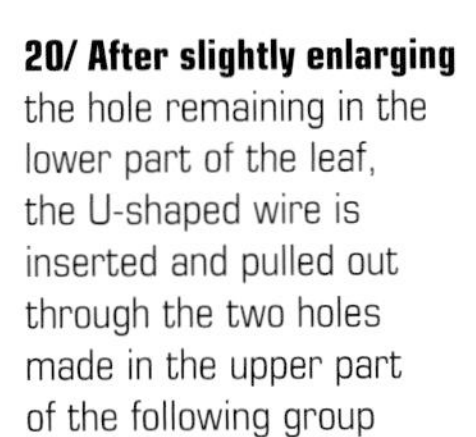

20/ After slightly enlarging the hole remaining in the lower part of the leaf, the U-shaped wire is inserted and pulled out through the two holes made in the upper part of the following group of leaves.

21/ The process is repeated for every group of leaves. Pliers are used to pull on the ends of the wires and bend them to secure all the leaf modules.

22/ The bracelet is placed on a charcoal block, and only one of the two wires of each joint is soldered. Once the gold has cooled and been pickled, only the soldered wire is cut off. Leaving one of the wires free creates the right tension in the piece so it will hang and move correctly.

23/ The bracelet is forced inward and lightly outward to produce the proper movement in the joints. The rest of the wires are soldered, and each end is filed and ground so all are fitted to the leaves.

24/ The clasp is prepared using two of the cast leaves. A tongue and latch are made from a laminated 0.05-mm sheet.

25/ This diagram shows how to make a box clasp.

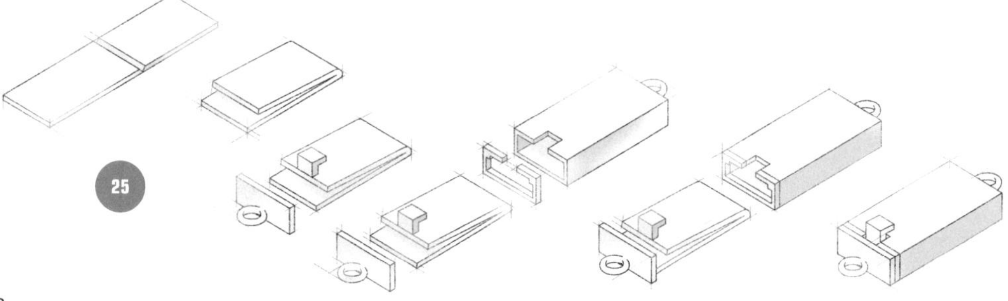

26/ The latch is soldered perpendicularly onto a 0.05-mm sheet so the tongue fits inside. Pliers are used to bend the back part of what will be the box for the tongue.

27/ This box-like structure is soldered to one of the cast leaves.

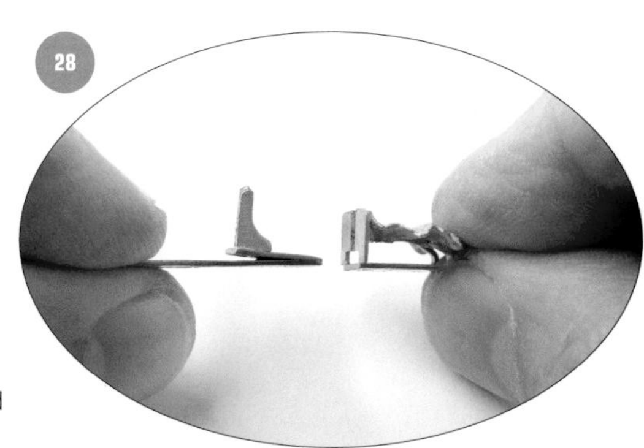

28/ A small, 1-mm sheet is soldered to the top part of the tongue in the manner of a push-release. The jeweler's saw is used to cut the opening for the push-release to fit inside the box.

29/ A latch is made in a U-shape to support the tongue, the end of which will be soldered onto the adjacent cast leaf.

30/ When the clasp is finished, a safety latch can be made if desired.

31/ Here is the finished bracelet after patinating it with an old-gold color.

Codina Goldsmiths.
Cast bracelet

Cast Orange-Peel Pendant (Walter Chen)

Walter Chen carefully selects unripe oranges and manipulates the skin of these fruits to produce a set of rings and a pendant cast in silver. This interesting project shows the many creative possibilities that can be produced from organic materials.

1/ Unripe oranges are carefully selected based on their size and the quality of their texture.

2/ It is important to have the right tools. These paring knives were modified for making efficient, precise cuts. One knife has a straight blade and the second one has a finer, curved blade.

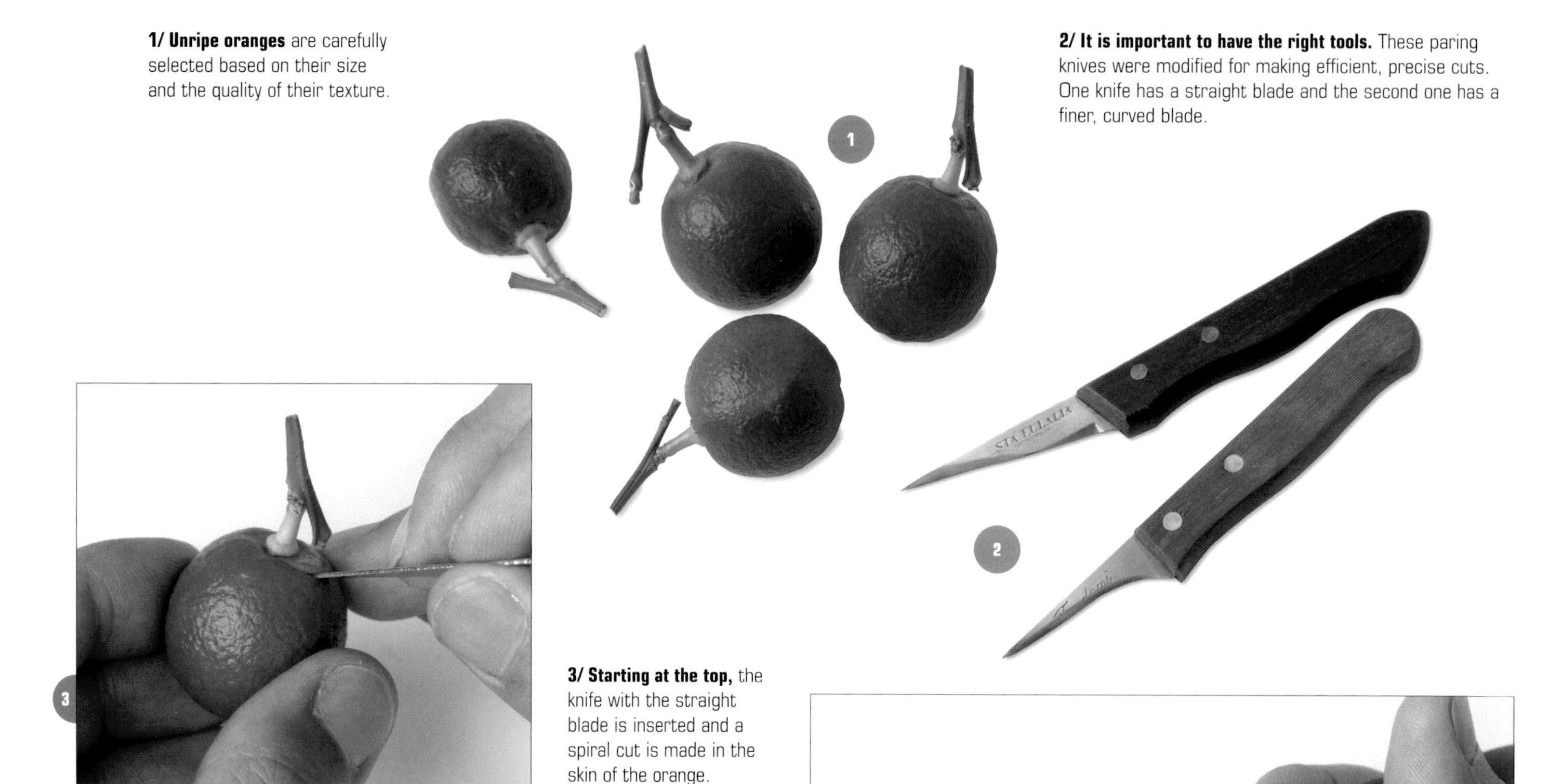

3/ Starting at the top, the knife with the straight blade is inserted and a spiral cut is made in the skin of the orange.

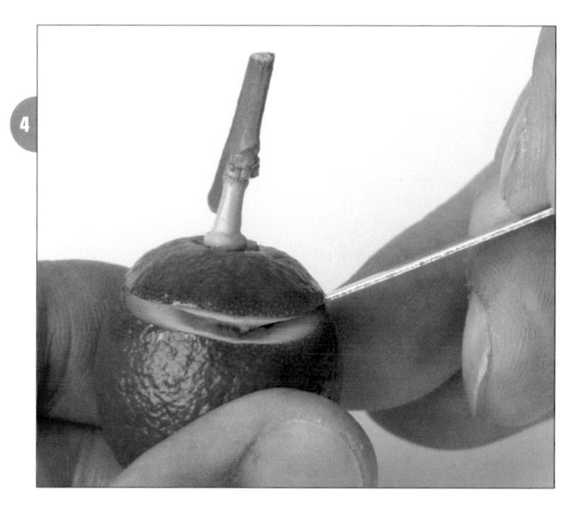

4/ The knife is inserted to a depth of about 5 mm to deepen the cut and to reach the core of the fruit.

5/ It is very important for the cut to flow in a regular fashion. The strip of orange should maintain a regular profile from beginning to end.

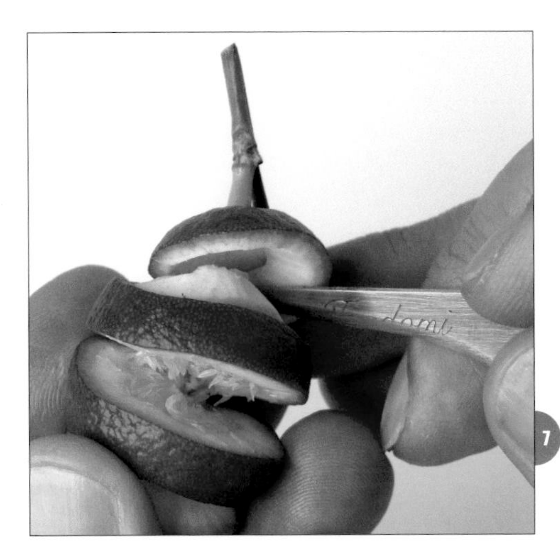

6/ When the cut is completed, the core of the orange is removed.

7/ The knife with the curved blade is used on the fleshy part of the orange to separate the harder outer peel from the inner core.

8/ A clean cut is made that yields a uniform strip of orange.

9/ The peel is dried before casting. A wood structure is made to keep the strip of peel separated and exposed to the air.

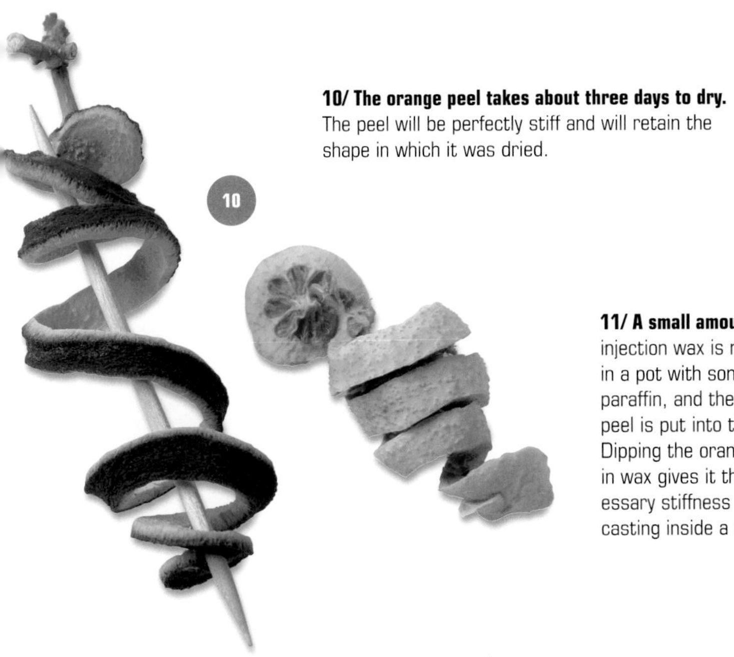

10/ The orange peel takes about three days to dry. The peel will be perfectly stiff and will retain the shape in which it was dried.

11/ A small amount of injection wax is melted in a pot with some paraffin, and the dried peel is put into the pot. Dipping the orange peel in wax gives it the necessary stiffness for casting inside a cylinder.

12/ Once the wax cools, the outside is brushed with a soft plastic brush. Remove as much wax as possible while always preserving the texture.

13/ Various runners are soldered on to the form to ensure proper feeding of the molten metal.

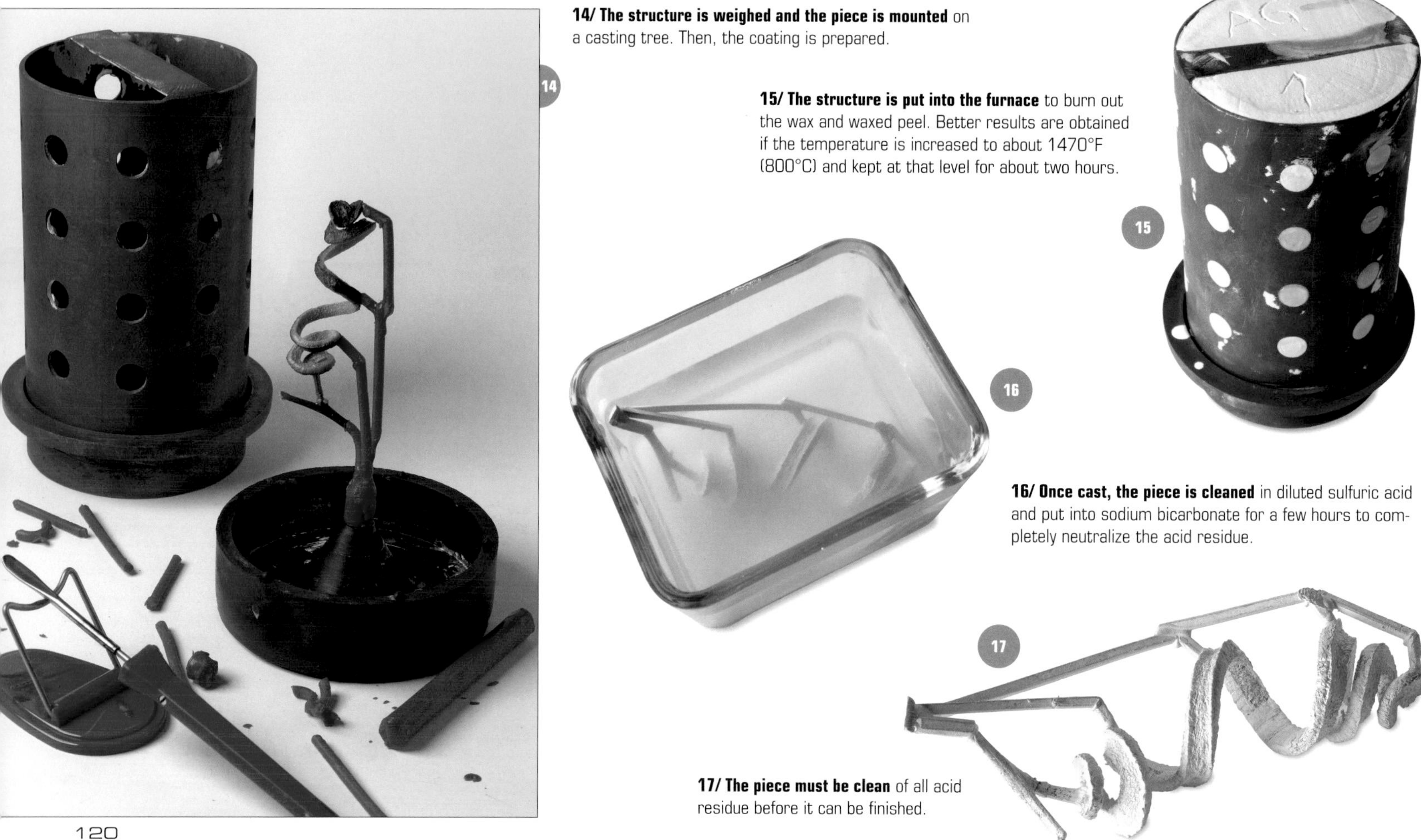

14/ The structure is weighed and the piece is mounted on a casting tree. Then, the coating is prepared.

15/ The structure is put into the furnace to burn out the wax and waxed peel. Better results are obtained if the temperature is increased to about 1470°F (800°C) and kept at that level for about two hours.

16/ Once cast, the piece is cleaned in diluted sulfuric acid and put into sodium bicarbonate for a few hours to completely neutralize the acid residue.

17/ The piece must be clean of all acid residue before it can be finished.

18/ A burnisher is an essential tool for finishing this type of piece. It is easy to make a burnisher from a piece of tempered high-quality steel.

19/ The outermost part of the texture is burnished by pressing it with a perfectly polished burnisher.

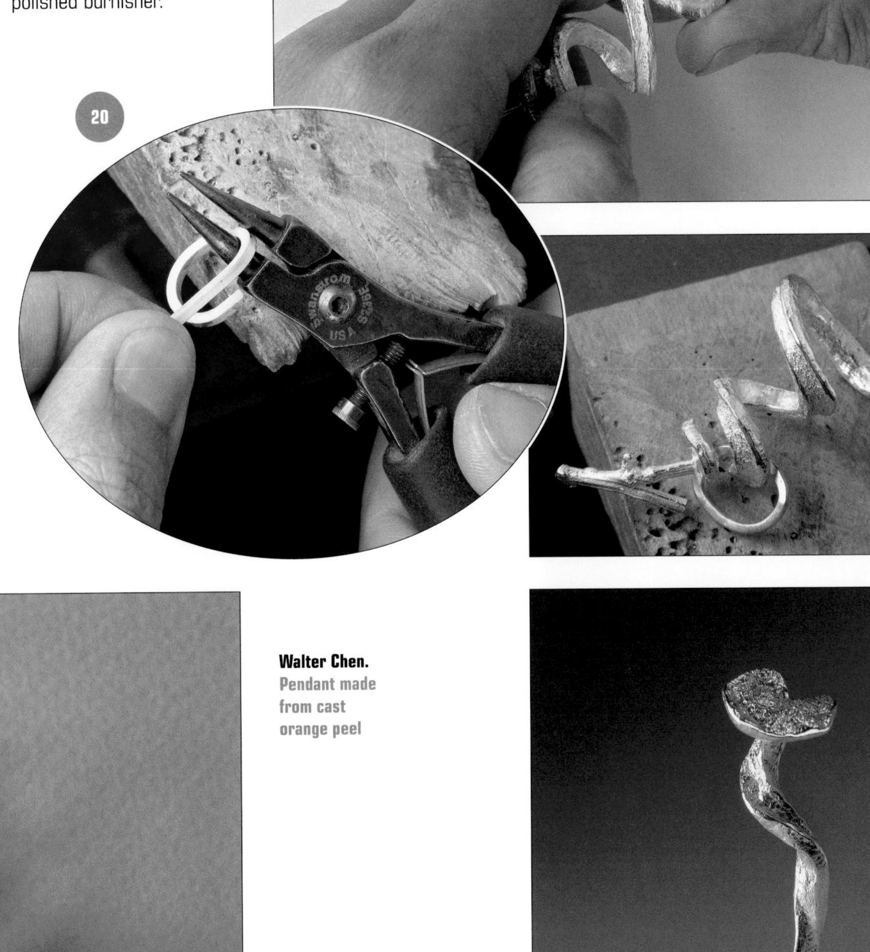

20/ A bale is made from square wire that is 1.5 mm thick so the piece can be worn as a pendant.

21/ The bale is soldered on, passing it right through the piece.

22/ The finished pendant

Walter Chen.
Pendant made from cast orange peel

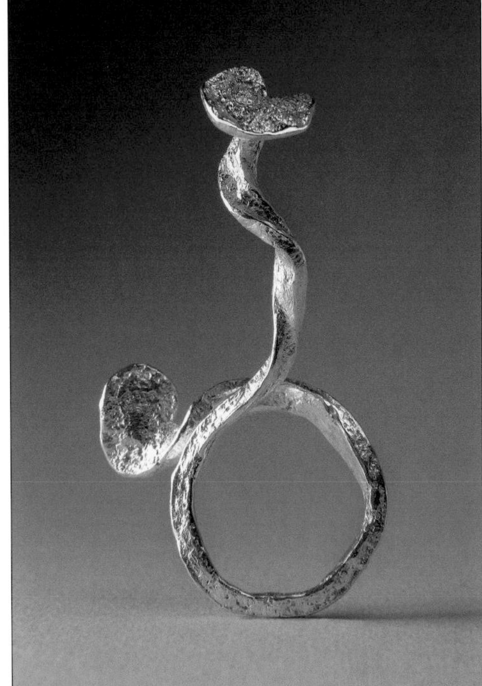

Walter Chen.
Ring made from cast orange peel

Pendant Cast From Polyurethane Foam (Daniel Fàbregues)

Polyurethane foam is used industrially as a filler and an insulator, but it can also be used in casting jewelry. It is extremely soft and fragile, very easy to cut and work with few tools, and it presents no problems when cast with precious metals. Since this is a porous plastic formed by a series of bubbles generated by a chemical reaction, this polyurethane takes on a very special rough texture that Daniel Fàbregues uses to create an original pendant in the steps below.

1/ To begin the project, we use a small piece of medium-density polyurethane foam and a precise design of the piece and its size.

2/ A box cutter is used to cut out the polyurethane, following a pattern previously cut from card stock.

3/ The outer edges of the foam can be filed or smoothed with emery paper on a flat surface.

4/ It is important to square up the outside of the foam frame perfectly. It will not be possible to remedy any imperfections once the foam is cast in silver.

5/ A craft knife is used to cut out the interior of the foam frame.

6/ The interior is filed with great care to create the desired profile.

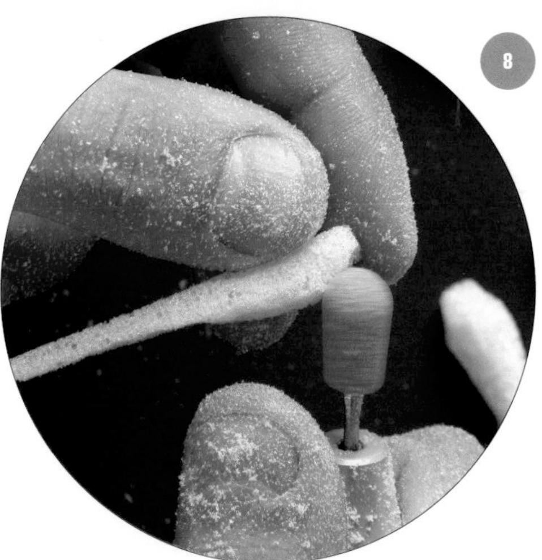

7/ The remaining components for the pendant are cut out directly with the box cutter. You must take the greatest care when working with very fine pieces, because this thin foam is very fragile.

8/ To finish defining the shape, you can use a burr like the one shown to cut the polyurethane.

9/ A sharp steel point is used to shape and correct the form and to create the desired profile and textural finish.

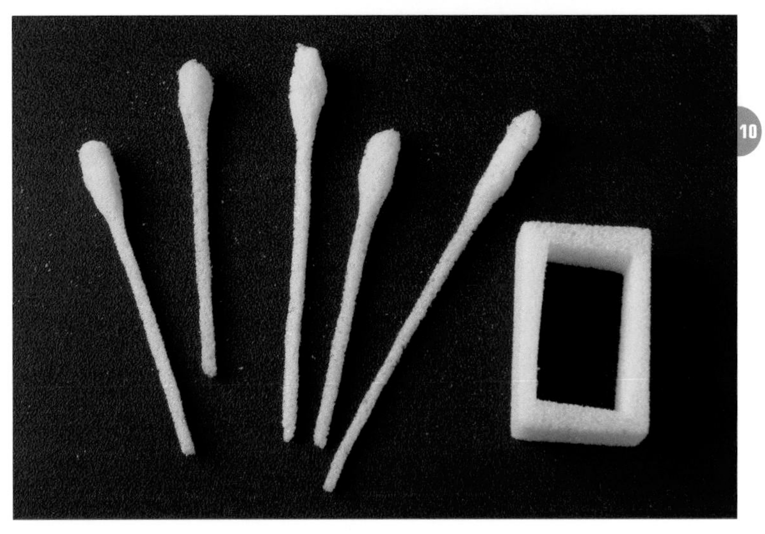

10/ Once all the elements are completed, they are weighed. Their weight is multiplied by the density of the polyurethane foam. This requires checking the foam manufacturer's technical data sheet.

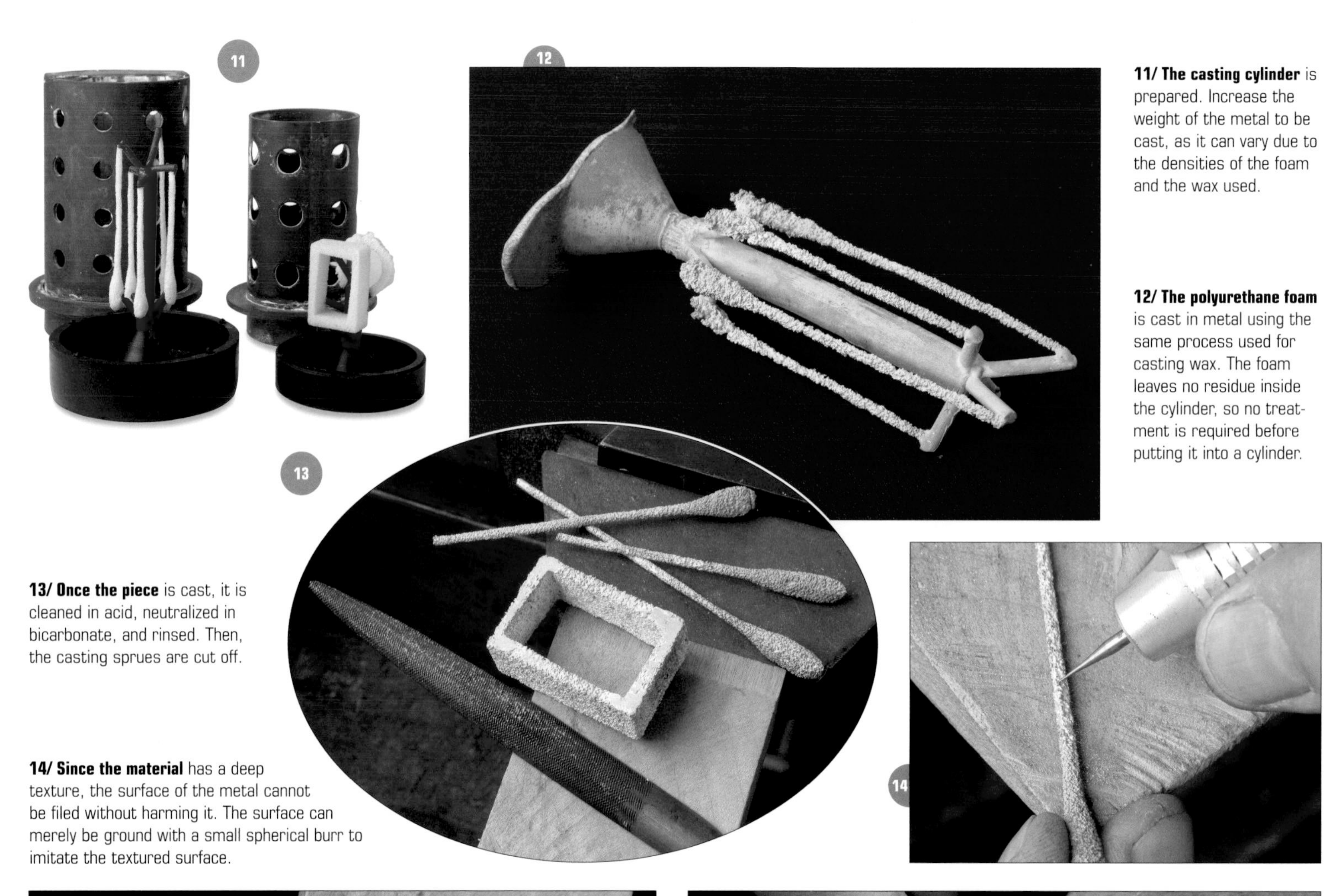

11/ The casting cylinder is prepared. Increase the weight of the metal to be cast, as it can vary due to the densities of the foam and the wax used.

12/ The polyurethane foam is cast in metal using the same process used for casting wax. The foam leaves no residue inside the cylinder, so no treatment is required before putting it into a cylinder.

13/ Once the piece is cast, it is cleaned in acid, neutralized in bicarbonate, and rinsed. Then, the casting sprues are cut off.

14/ Since the material has a deep texture, the surface of the metal cannot be filed without harming it. The surface can merely be ground with a small spherical burr to imitate the textured surface.

15/ A hole is drilled in the upper part of the pendant so a steel wire can be passed through, making it wearable.

16/ A larger drill bit is used to make several transverse holes to accommodate the cast silver pins.

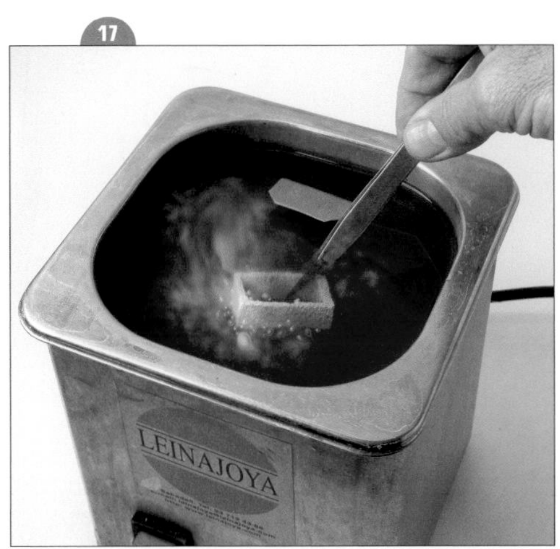

17/ To make the silver very white, the frame is oxidized with a torch and dipped into pickle. This process must be repeated until the spots are eliminated and a uniform white color is produced. The frame is completely cleaned.

18/ The pins are oxidized with silver oxide, and then coated with metal varnish. This process improves the shine and resistance to oxidation.

19/ Each pin is fit into place. To adjust the depth of the pins, modify the holes in the frame

20/ The finished piece with one of the pins plated in 18-karat gold

Daniel Fàbregues. Pendant made using polyurethane foam

Daniel Fàbregues. Pendant made of silver and oxidized silver

Acknowledgments

Many generous professionals collaborated and contributed images of their work, which was essential in sharing and teaching the procedures and techniques in the first part of this book. Our sincere thanks to Miquel Gasso, Akiko Kurihara, Marta Sánchez, Michael Zobel, Andronikos Sagiannos, Simon Victoria, Dani Fàbregues, Harold O'Connor, Carmen Amador, and Stefano Marchetti. Also to Clara Inés Arana, who first taught me the keum boo technique. And to all the authors whose works, for reasons of space, could not be reproduced in this volume. We are especially grateful to Tensi Solsona for his collaboration and for contributing his unique granulation procedure, and to Hans Leicht, who is a companion and collaborator in this volume as well as a jewelry professor at the Escola Massana.

The second part of this book, which focuses on casting, required a tremendous amount of work because of the variety of techniques presented. The collaboration of Walter Chen, who shows how to cast fine, delicate organic elements, helped provide the proper understanding; this also applies to the invaluable contribution of Tensi Solsona, who shows how to create rings right in the casting sand. We also wish to thank Daniel Fàbregues for making a pendant directly from polyurethane foam.

We also thank Joan Codina for the creation and prototyping of computer files, and the generous contribution of his work, which is essential to this volume, done by Elisa Pellacani and Carolina Hornauer, plus various photos of procedures kindly provided by Rafael del Molino and Àlex Antich.

Many thanks to all,

Carles Codina i Armengol

Index